Raised Bed Gardening – <u>5 Books bundle</u> on Growing Vegetables In Raised Beds & Containers (Updated)

<u>By</u>

<u>James Paris</u>

Published By

www.deanburnpublications.com

2nd Edition Updated March 2015

ISBN-10: 1490484035
ISBN-13: 978-1490484037

Blog: <u>www.planterspost.com</u>

Gardening Books Included In This Bundle

Book 1: Raised Bed Gardening - Ten Good Reasons For Growing Vegetables In A Raised Bed Garden

Book 2: Raised Bed Gardening Planting Guide – Growing Vegetables The Easy Way

Book 3: Raised Bed Gardening Planting Guide (2) - Making The Most Of A Raised Bed Garden For Growing Vegetables

Book 4: Vegetable Container Gardening – Growing Vegetables In Containers & Planters

Book 5: Tomato Container Gardening: Growing Tomatoes In Containers, Planters & Other Small Spaces

Relevant Books by Same Author

Raised Bed Gardening 3 Book Bundle

Companion Planting

Square Foot Gardening

Square Foot Vs Raised Bed Gardening

Vegetable Gardening Basics

Companion Planting

Straw Bale Gardening

Subjects covered include many no-dig gardening techniques for growing vegetables in limited spaces or small space gardens, such as Raised Bed Gardens & Container Gardens. Growing tomatoes in containers. Composting & creating 'Home Grown' compost.

Urban gardening & organic gardening methods including natural pest control. Growing vegetables indoors, companion planting methods, organic tea, potato planters & many other techniques for growing vegetables in small spaces or even city apartments.

Copyright

Table of Contents

Contents

Book 1

Raised Bed Gardening - Ten Good Reasons For Growing Vegetables In A Raised Bed Garden

By

James Paris

Introduction

I know quite a few people that have considered the aspect of growing plants in a raised bed, but consider it to be a lot of hassle and very expensive. This is far from the truth, and in fact growing anything in a raised bed is cheaper by far in the long run, than growing in a traditional garden setting. You may have noticed here that I am relating to planting in general terms, and not specifically referring to vegetable planting as is the usual case.

The reason for this is quite simple. A raised bed garden is not just for vegetable growers, but in fact covers the full spectrum of growing in general including, fruit growing, herb growing, growing flowers and of course that for which it is most commonly associated, growing vegetables.

In my previous books on Raised bed gardening, I have covered the main aspects of growing in a raised bed. However I have only touched on the reasons why you should perhaps give it a bit more thought – if you have not already done that. So in essence this book is not so much **how-to**, as it is **why-to** grow in a raised bed situation.

As you may have guessed by now, I am quite a fan of raised bed gardening, as it is simply the best way to go when it comes to production versus time spent. Personally I am into vegetable gardening, so the slant of this work will naturally head in that direction. However bear in mind that just about all the aspects covered here will apply to flower or fruit growing just as easily.

So what are the advantages of this type of gardening over the traditional garden plot? The following chapters should explain all. In **no particular order** then, here are the..

Ten Good Reasons Why…

1: Ease of Operation: When it comes to using a raised bed for growing, then the fact is that most of the work has already been done, with the construction and filling-in of the structure itself. This is of course covered in my earlier works.

The managing of the bed itself, is far and away easier than the traditional gardening method of digging out a garden area and then planting out in long rows – as is the case with vegetables for instance.

I should perhaps point out here that when I say 'traditional', this is not the case for many countries who have adopted the raised bed method, going back for centuries. Mountainous regions in particular have adopted a terraced approach to farming that is not dissimilar to raised bed gardening – as we shall see in a later chapter.

Extreme Example of Terraced Bed!

So what makes it easier to operate? For starters the ground in a raised bed garden does not get walked upon, as does the ground when gardening in the traditional way. This means that the soil (compost) is not all compressed, and difficult to turnover or weed for instance.

Because a typical construction is only 3 to 4 feet across, then there is really no need to be walking and thereby compressing the growing area. If this is necessary for whatever reason, then it is a simple matter to lay a plank across and walk on the plank rather than the bed.

In fact, laying a plank across the width of the bed that you can shift around is a handy thing to do, as you can use this as a kind of mini workbench on which to lay your garden tools – or even a glass of something cool to 'wet your whistle'.

If you build your raised bed with an edging plank around it; this makes for a handy seat from which to tend your garden area, with minimum effort

2: Best Use Of Space: A raised bed garden is particularly suited to those with a limited space in which to garden, although not exclusively so, as I will explain.

If you have a small area in which to plant your vegetables, even as little as a 10 x7 foot area, then you can build a raised bed 6 x 3 in size. This will give you a two foot space all around the raised bed in which to tend to the plants with ease. Although it is obvious that the ground for the traditional garden area is larger, the fact is that more vegetables can be grown in the smaller space owing to the different planting techniques.

When planting in a raised bed, you are planting using the 'in row' rather than the 'between row' method. This means that the planting in a raised bed is much more intensive and makes best use of the area available. This is mainly down to the fact that you do not have all the wasted space in walkways, that you have in a traditional garden, in order to care for your plants.

It has been estimated that where a traditional garden wastes up to 68% of the actual ground area. The raised bed system only takes up around 37%. This means that the raised bed is making full use of around 63% of the area available to actually produce something, compared to a traditional gardens 32%. Granted, this is not an 'exact science' but I think give a fair idea of the way things are.

If however you have a larger area to plant in, this is where a raised bed system becomes even more effective. The reason being the walkway around the beds, now serves two beds instead of one for the same planting area. It is simply the 'economy of scale' model in action. Where one bed fits in an area 10 x 7 foot. Two beds will fit in an area not double the size, but rather 10 x 12 foot. I'm sure you get the general idea!

If you then take in the best use of the space available, then it is obvious that the raised bed comes up trumps in this department.

3: More Accessible: Another major advantage of the raised bed system, is simply the accessibility of the thing. I personally suffer from a bad back, owing to all the heavy work I've done over the years as a builder. Consequently, after just a short time bending over pulling out weeds, or even harvesting and tending the vegetables, my back is killing me!

The solution is simple and very effective, build a raised bed garden about 18 inches high. This makes it a perfect height to manage, without all the bending over that you have with the traditional garden method. Not only that, but the seat you create around the structure means that you can actually tend your garden whilst sitting down !

On another level; if you have a disability or impairment of any sort that makes it difficult if not impossible for you to tend a garden; then a raised bed system might just make it all possible. Make the lanes between the beds wide enough for a wheelchair – if that is needed, making sure you have a sound level surface to move on, and you are good to go.

This system of raised bed gardening has given many wheelchair-bound individuals a whole new lease of life, especially in communities where the local authorities have taken the idea on board.

4: Less Weeding!: Weeding the garden must go down as one of the less enjoyable of garden chores, though I have heard that it can be therapeutic in its own way, hmmm. With this system of gardening the whole weeding business is far, far less of a task.

The main reason for this is simply the fact that you are not planting in a weed seed infested piece of ground; but rather in an area of virgin ground that you have provided specifically for the job in hand. I am referring of course to the fact that when you filled in your raised bed structure, you did not fill it in with soil from the garden or elsewhere (at least I hope not !).

The infill of a raised bed changes slightly according to the crop (or crops) you wish to grow, but basically it should be made up of a good compost mixture including plenty of organic material such as well-rotted manure for instance. This can be mixed with some garden soil, but no more than about 20% soil to 80% compost. This will help with two things; first it will prevent the soil from compacting in the raised bed, and second it will mean that you are not importing the weed seeds into the raised bed area.

This all results in less weeding for you to do, and more time to do the things that are of greater importance, like harvesting some fresh peas for the evening meal, or plucking some fresh tomatoes for that crisp summer salad!

Does this mean that there is **no weeding at all** in a raised bed garden? Simple answer is no, it does not. However this time-sucking chore is dramatically cut down by these facts - The soil is looser, making the weeds simple to pull up. There are fewer weeds to begin with, owing to the new compost you have filled in with.

5: Free From Garden Pests: OK, I'll be honest; when I say free from garden pests, it is a bit of an exaggeration – or do I call it artistic licence?! Whatever, in my defence, it is definitely much easier to keep pests of all persuasions out of a raised bed garden than it is to keep them out of a traditional garden plot.

For instance, if you have followed my guidelines in my previous books for keeping out burrowing creatures such as Gophers, Moles and mice; you should not have any trouble at all with these creatures. The whole idea of a wire mesh barrier at the bottom of the raised bed works every time.

The greatest pest for me is the rabbit, and the damage that one rabbit can cause overnight to my vegetables, means that in my book they are public enemy no1. Here again the raised bed comes up trumps, because rabbits will seldom jump up over 18 inches to two feet – making it easy to keep them away from the vegetables with a simple barrier to raise the sides of the structure if need be. Since they cannot burrow under this barrier, as they are prone to do with a fence, this is very effective.

On another level, for instance with the carrot fly, you already have a huge advantage if you bed is built around the 18 inches high level. The reason is that the carrot fly does not tend to fly above 2 feet or so, which means that a simple extension around your bed with a simple fly mesh to make it over two feet high, means that you will avoid the worst of these pests.

When it comes to such things as mildew and overly damp situations, a raised bed has the advantage of catching any little breeze that is going around. This all helps with the internal ventilation between the plants, the lack of which is the great cause of many diseases amongst things like tomato plants for instance.

With flying pests like birds of moths, again it is a fairly straight-forward matter to erect a frame attached to the structure, that will allow you to install over the frame some nylon mesh. If done with some fore-thought, this will allow you easy access to your plants without all the drama of getting mesh tangled up in the plants, that you can have in a traditional garden area.

The actual removal of or treatment for pests is also much easier with a raised bed, owing to the fact that you are not bent over double trying to work out what the problem is – it's usually there staring you right in the face!

One of the top tips to help avoid insect infeststation is crop rotation. This can be done easily with a system of raised beds, and is a sure way to help avoid contamination through the over use of a particular growing area or plot of ground.

6: Longer Growing Season: Another often overlooked plus of the raised bed system of growing, is the longer growing season that you have with this type of gardening. Because the compost is lighter in make-up and is raised up away from the ground, it means that it warms up quicker come the springtime. Meaning of course that you can have a longer growing period, simply because you can plant that much earlier. If you couple that with some of my tips in the earlier books then you can extend this even further and grow much longer into the season that you normally could.

Of course a longer growing season, means that you can potentially have a much larger harvest. This in turn has an immediate effect on your costing's overall, and can transform your growing enterprise into a real money-maker, helping to balance the monthly food bills. This in turn helps to offset and even remove over time the initial cost of setting up the whole material costs of your garden project, from the bed construction down to the cost of plants and materials.

7: Perfect For An Uneven Surface: As I alluded to in the first chapter, raised bed gardening is nothing new, and in fact goes as far back as the Incas and even beyond that. The fact that a raised bed system can be used to make an area of ground that is unproductive because it has too much of a slope on it, was not missed on the Incas.

Yes you may well say that this was more of a terraced gardening regime, I would answer however that even the terraced garden can be similar to the raised bed, inasmuch as it seeks to level out a piece of ground for growing purposes.

Even today you will see this system all over the world, particularly in mountainous regions such as Nepal or the South Americas. By utilising the raised bed or terraced system whole villages in these mountainous regions are able to provide all the vegetables that they need.

OK, so you may not live in Nepal! However maybe you have a garden plot that is on the side of a hill, and you have always considered it unsuitable to grow anything? The answer is a raise bed system, that has the long side broadways on to the hill. Make one side higher than the other to compensate for the slope. For instance one long side may be just 6 inches high, where the other is two feet high; if this is what it takes to level the growing surface so be it.

In some case it may work better to narrow the width of the bed to say about two feet instead of three. This would be the case in the steeper situations, as it reduces the height of the high side of the bed. One thing to note when building on a hillside, is to be sure that your frame is well anchored to the ground with posts or even iron stakes. The last thing you want is for gravity to do its 'thing' and carry your vegetable plot down to the bottom of the hill !

8: Easily Modified: A raised bed must be one of the most versatile of gardening techniques, because it is so easily modified, and adaptable there-fore to a wide range of growing conditions or demands. For instance with a simple structure applied with one inch plastic tubing and some clear polythene, you can have a min-greenhouse in which you can grow fruit or vegetables that may otherwise need a hot-house, depending on where you live.

This same structure can also be used to throw a nylon mesh over for bird or insect protection; if for example your 'greenhouse effect' was only to last you over the cold spells of early springtime. Another good use of the raised bed is to add a framework, enabling you to grow all sorts of climbing plants such as pea or bean crops. You may well say that you can do this with a traditional garden also? This is true, however it is just that much easier and simpler to build the frame and manage the crop, within a raised bed system

Particularly if like most people you have built the bed out of lumber, then it is a simple matter to attach whatever growing frame you have to the overall structure of the raised bed. If you are care full to attach this frame with screw-nails rather than just ordinary nails, then it is easy to remove this in order to meet your future growing needs, without damaging the frame itself.

9: Great For Diversity: It is hard to discuss a raised bed system of growing, without also mentioning – at least in the passing – the concept of Square Foot Gardening. This is a system of gardening, where every crop is grown in an area of one square foot, and whilst this may seem silly if you have not come across it before; it can actually work fairly well if you get a grasp of the system.

The basic precept is that you start with a four foot by four foot frame-work, which you then divide into one foot squares by string or some other simple narrow frame. This then gives you 16 squares in which to plant your various vegetables. The theory is that, provided you rotate the planting with regard to space time and nutrient demand; then this system is able to supply a family of four with vegetables throughout the whole of the growing season.

It is of course a little more detailed than that, but that is the 'bare bones' of the Square foot gardening system. To go into the ins and outs of it, I would need to write another book..hmmm, there's a thought! Anyhow, it does not take too big a stretch of the imagination to see that a raised bed is very adaptable to this kind of system.

Another aspect of diversity is particularly relevant if you have more than one raised bed. This enables you to concentrate on just one or maybe two crops per raised bed, meaning in turn that you can prepare the compost or infill, to meet the particular demands of that crop. For instance if you are growing carrots or parsnips, then you would make sure that the compost is light and loamy with plenty of sand added to ensure the best crop.

Like-wise for crops like leeks, that needed plenty of well-rotted manure in order to thrive; can be well catered for in their own individual raised bed garden.

10: And Finally: No More double digging! This was always one of the most laborious tasks at the beginning of the season, double digging in order to get some 'looseness' back into the solid ground. This was made solid by constant walking up and down between the rows of growing vegetables; now however with the raised bed gardening system, double digging is a thing of the past as the compost can easily be turned over with a small garden fork.

You may well ask, but what about the soil getting stale? This would be a problem if you continuously grow the same crop in your raised bed situation. However if you rotate the crops properly, either between beds or within beds, then the compost needs little work or additives to keep it healthy and 'producing the goods' so to speak.

One of the aspects of double digging is that it turns up soil that is still 'virgin' at the bottom and puts it nearer the top growing area. However if not done properly double digging can result in the subsoil being mixed with the topsoil, which is disastrous for the plants.

A raised bed system avoids this situation entirely by removing the need to double dig. The compost has already been built up with plenty of organic material, allowing for aeration of the soil and the better distribution of nutrients. And since a raised bed is not prone to compression in the way that a traditional garden is, there is no need to drastically loosen up the growing medium.

This in effect cuts out entirely the need for double digging a garden area.

Summary

I did mention at the beginning of this work, that I am a fan of raised bed gardening! When thinking about this book, I had intended to name it 'The pros and cons of a raised bed garden' however I soon realized that the trouble with that title, was that I could not find enough cons to justify it!

The fact is that there are very few cons to raised bed gardening, if you put aside the early effort and expense to build and fill the thing up. However there is time and effort in everything that you do, so even that is not a fair appraisal.

As for the money or financial commitment, then that may seem a 'con' until you add up the benefits re more productivity, and time not wasted elsewhere – time is money after all.

MY NOTES / TO-DO PAGE

MY NOTES / TO-DO PAGE

MY NOTES / TO-DO PAGE

Book 2

Raised Bed Gardening Planting Guide – Growing Vegetables The Easy Way

BY
James Paris

Introduction to Raised Bed Gardening

"The love of gardening is a seed once sown that never dies." -- Gertrude Jekyll

You may well be wondering why on earth a raised bed garden is any easier than planting vegetables straight into the soil. Or why indeed I have titled this as I have done, by insinuating that raised bed gardening is easy.

Well the fact is that in my opinion, growing vegetables in a raised bed is by far the easiest way of growing great vegetables without the huge labour involved when growing the traditional way.

However I must clarify that by saying raised bed gardening has been around since the beginning of time, and although it has received more prominence as of late, it is by no means a new concept – think of the hanging gardens of Babylon !

Nothing however is without its cost, in both labour and financial expenditure, and it is true that to grow vegetables in a raised bed requires an element of both, especially at the early stages. This however need not put anyone off the idea, as the effort is minimal and short lived compared to the benefits derived from a raised bed garden that 'produces the goods'.

In this publication I intend to spell out, in layman's terms, just what it means to grow vegetables (or fruit, flowers etc) in a raised bed. How to construct a raised bed, simply and easily, including the different materials that can be used at minimal cost wherever possible. Even how to convert your raised bed into a temporary greenhouse, at minimum
cost in easy to follow steps.

The raised bed gardener is able to spend longer tending to his plants than the average vegetable gardener, simply because he (or she) is not spending their valuable time digging over soil and clearing out weeds. For this reason alone the raised bed is preferable for those who are working all day and have limited time to spend in the evening tending their vegetable plot.

Advantages of a raised bed garden

There are several advantages that a raised bed has over planting straight into the ground, some of these are as follows:

With a raised bed, it really does not matter what quality your garden soil is, or indeed what the drainage is like. As this is all added when forming your raised bed garden.

Easy to service/maintain. With a raised bed you have the advantage of height, which means that you do not have to bend over as far to take care of your vegetables. This is particularly advantageous if you are prone to suffer from back-ache.

Weed free. A raised bed is not troubled to nearly the same extent by the incursion of weeds, as all the soil/compost mix is freshly added. For any weeds that do appear, they are easier to remove as the compost mix does not compact like garden soil.

It is far easier to control destructive pests within a raised bed garden. This is simply because you are off the ground, and so keeping a natural barrier up in front of creeping pests like garden slugs.

With a slightly higher raised bed of around two feet, then you are not troubled quite as much with carrot fly for instance, who tend to be low fliers.

So out with back-breaking weeding tasks, along with digging over water logged soil and filtering out rocks and stones. In with easy gardening methods for the busy householder, and fresh vegetables for the whole family with the minimal of hassle.

Building A Raised Bed Garden

"When the world wearies and society fails to satisfy, there is always the garden."-- Minnie Aumonier

A raised bed can be built with any one of a number of materials, the most popular being timber (untreated). Other materials include, concrete, brick, corrugated sheet metal or concrete block work. In fact anything that you have to hand that can produce a decent barrier about 1 foot to 18 inches high, can be used for a raised bed.

Raised Bed Dimensions:

As to dimensions, this is really determined by many things including the space you have available, and indeed just exactly what your requirements are. Do you have a large family to feed, or do you intend to sell or barter (bartering is a great way to enjoy a diversity of produce from other gardeners) some of your produce ?

With all that considered, a typical raised bed vegetable plot is about 6 foot by 3 foot. This is an ideal size because it allows access from both sides, without you having to step onto the raised bed itself. This enables you to tend to your plants without treading on them in the process – always a good thing !

As to depth, overall you should aim for 1 foot minimum depth, up to two feet for deep rooted plants such as carrots for instance.

The depth of the raised bed does not have to be the height of the sides, to explain a bit further. Say you would like a bed depth of 18 inches (450mm), but you only have timber for 12 inch sides. Simply build your bed in the manner described later, and dig out the interior an extra six inches. This will enable you to fill in the bed with the compost of choice up to the required depth.

It should perhaps be pointed out here though, that this negates the concept of building a raised bed for the advantages to be gained with the height of the bed itself above ground, as will be explained later in the article. This system is mainly used where the existing soil is of poor quality and has to be replaced/substituted in order to grow the vegetables of your choice.

If you are building multiple raised beds, then they should be placed about two feet (600mm) apart if possible, to allow for easy access between them.

Building a timber raised bed

The construction of a timber raised bed is fairly simple and straight forward. First of all, level and mark out the area where you would like your raised bed to be. Bear in mind that it should not be under overhanging trees, and in an area where you can have easy access for tending your plants. It should get a minimum of 5-6 hours sunshine per day to produce best results for most vegetables.

For a 6 x 3 x 1.5 foot bed built using traditional decking timber, you will need:

6 lengths decking @ 6' x 6" x 1"
6 lengths decking @ 2'10" x 6" x 1"
10 – 3" x 2" pointed posts @ 30"
Weed control fabric
Galvanized screws or nails
Wire mesh (optional)

Begin by marking out with string and pegs, the area of your raised bed, putting down a peg on each corner. This is where you should consider whether or not you are going to dig out any of the existing ground.

Questions to ask yourself are, what depth of compost do I need, versus what height do I want the finished bed to be. If you are growing root vegetables that need depth, but you do not want the finished height to be over 1 foot for instance, then digging out the area to the depth required is your only option.

Once this decision is made, then we can proceed with building the raised bed. Once you have the pegs in the area that marks out the four corners of your raised bed, you simply take out one peg at a time and replace by hammering down your pointed posts, leaving them a minimum of 18 inches above the ground.

Alternatively, if you make these posts longer then you can use them as handy aids for lifting yourself up when tending your vegetables – just a matter of choice really.

The best way to do this is to put down one post at the end, then temporarily fix the first short end against the post. With this done, then hammer in the second post flush with the end of the 6" x 2" decking plank. Proceed with the two longer sides, then complete the other end. If you just put one screw partially home, then you can easily adjust to suit.

Be sure that you have leveled the timber and that you have left a minimum 12" in height above the first planks, so you are able to complete the job.
I find that it is better to construct with a cordless screwdriver as this does not impact the framework in the same way that hammer

and nails does. Also should you make a slight error, then it is no trouble to take apart for adjustment.

Once this is done then simply mark out along the inside length two feet from each end, then making sure the construction is straight, hammer in two of the posts to the same height as the others. On the end of the construction, do the same with one post in the centre of the framework.

This will give you a strong sturdy construction, which you will need if you do not want the sides of your raised deck to bow under the pressure of the soil.

Point of note:

If you are building with heavier timbers, say 6" x 2" for instance then it may be possible to just put one post in the center of the long side and none at all on the end. I however tend to lean on the cautious side, and would rather aim for stronger option overall. Another tip is to put a cross brace in, if you are concerned about the sides bowing outward.
It is not an exact science, but there are minimum guidelines that must be kept to ensure a construction fit for purpose.

After you have built the sides then just screw down the remaining planking face down along the edge (as in the photograph), to make a comfortable sitting or leaning area for tending to your raised bed.

One thing to consider during this time, is whether or not you are bothered by Gophers or Moles. If you are, then at this point you would place in 1" galvanized wire mesh, covering the bottom of your raised bed. This will be extremely effective in stopping the varmints from destroying your crop and giving you endless grief and heartache!

The weed control fabric should be fixed down the inside of the bed, to keep the wet soil away from the timber. This will help the timber to breathe and make it just that bit longer lasting.

2nd Point of note: Do not use timber that has been treated with creosote, as this may weep through and kill the plants!

Building a raised bed from brick

If you are fully convinced that the position of your raised bed is permanent or that you will never need to move it to somewhere else; then the option of building a solid brick construction is open to you.

The advantages of a brick-built raised bed are simply that it will not rot, and is a sound structure not easily damaged. Disadvantages are that of course, you cannot move it to another position without destroying it, and also it can look quite unsightly depending on the surroundings. If for instance you have a brick-walled garden, then it may blend in very nicely.

Materials that you will need for your brick-built raised bed are as follows, based on a 6 x 3 construction:

Approximately 400 common brick
Sand and cement
Crushed gravel
Concrete ballast

With any brick construction then you have to build a foundation, otherwise the construction will crack as it subsides into the ground. This is quite a simple construction overall, but you may feel more comfortable getting a builder to do it if you have no experience at this sort of thing. However if you are at all interested, I would suggest that it is a good time to try out your building skills!

Begin by marking out and leveling the ground, as in the preceding chapter. When this is done, then dig out a trench for your foundation, bearing in mind that the brickwork has to be roughly centre of the foundation. The trench should be about 12 inches deep (for frost cover) and 12-18 inches wide, if you are building with 4" brick. This allows for a good concrete raft to build upon. If you are building in an area not bothered by penetrating frost then this trench can be shallower – or deeper if you have the opposite problem.

Mix up a concrete mix with your concrete ballast using a 5 – 1 mixture. That is 5 parts ballast to 1 part concrete. Add water and mix thoroughly. Fill the bottom of the trench a minimum of 3 inches deep with this concrete mix. This will ensure a good solid foundation for your construction.
In just a couple of days this will be dry enough to proceed with the building process.

Top Tip:
If you want to shortcut this process a little, then for this construction it is possible to simply lay 4" concrete block on a bed of cement mix, to form a ready-made foundation for your walls.

Once the foundation has been laid then it is time to put your building skills to the test. Main thing here if you have no experience with building is to string a line along the edge of your construction, and follow it. **Keep the spirit level at work** and be sure not to deviate from the line.

Mix up your sand and cement to a 3 to 1 mix. Three parts sand to one part cement. Add a plasticizer to the water before using as this will make things a lot easier when it comes to spreading the compo. Cement mix should be about ½ inch deep and the same at the ends. Each brick must be laid level, with a slight tap to bed-in. Be sure to overlap each brick as you are building, and tie in at the corners.

On the first layer to rise above the ground level, you should include a few 'weep holes'. This is simply done by keeping a couple of bricks without compo at the joints and leaving a gap instead.

Once you have reached the full height, you can either just finish off with a bed of cement over the top bricks, smoothing to form a half curve. Or you can finish with a concrete coping from the builders suppliers.

Dry stone raised bed

Actually made from concrete block, laid flat; this is a simple construction that can be taken down when and if, it's not needed any longer.

If you use 18" x 9" x 4" dense block, then layout a flat area for the base, pounding in some crushed rock for a foundation. After making sure your foundation track is perfectly level, using a straight edge; Start to lay your block on the flat side down on a bed of rough sand.

This row must be perfectly level otherwise you will face problems as the structure rises. Make sure that you overlap the blocks so that there is no break going up through the wall.

The down side with this raised bed is that you will use twice the concrete block as building normally, however you will save on sand and cement as well as time.

Drywall Example Above

Finished result should be a solid construction that has a good broad top to sit on while working your raised bed. True, it takes up a bit more space, but overall it is perhaps the simplest and quickest way to build. Just be sure of the first layer, and everything else will follow on.
Be sure that you tie in the corners using the same building method.

Top Tip: If you would like a more secure finish, then simply lay the top row of block on a bed of cement mortar. This will secure the whole structure quite nicely

Other Raised Bed Examples

There is actually no limit to the amount of ways to construct a raised bed garden area, or indeed the different materials that can be used for it. Or perhaps I should say that the only real limit is your imagination!

Corrugated iron sheeting, properly supported is often used to create a raised bed. It has to be said though that if you are building for appearance, then this is probably not the one for you !

Timber logs cut straight from the tree. These can look especially effective and can be built similar to a log cabin construction, giving an extremely strong and versatile structure that will last for many years.

Old Railway sleepers. I would not particularly recommend using old railway sleepers, as there is a danger of creosote leaking into your plant bed, causing a health hazard – as well as killing the plants. If old sleepers are used then be sure to line the inside with polythene barrier to prevent this happening.

In general however the **modern railway sleepers** for sale in your local garden centre will not have been treated with creosote, but with a plant-friendly injection treatment. This makes them ideal for raised bed construction. Rot – resistant cedar or redwood are the best railway ties for building your raised bed. Consult the sales person before purchasing.

Build using the same principles above for the timber raised bed, but because of the heavy timber (about 19" x 5") you need only use support at the corners, except for the really long lengths at over 3 meters.

Filling and Modifications

*Let my words, like vegetables, be tender and sweet, for tomorrow I may have to eat them. - **Author Unknown***

Filling the raised bed:
Next we come to filling your raised bed. This starts with the drainage at the base. This can be broken pots or rough broken brickwork, built to about three inches deep. However before this you may want to consider whether or not you are bothered by moles, gophers, voles etc.
These creatures are likely to follow the worms or fresh shoots up into your raised bed if you allow it to happen.

If you are in any doubt then lay out your 1" galvanized wire mesh at the base of the bed, before putting in your drainage level – better safe than sorry ! Also a layer of tough weed suppressant material at this stage, laid over the mesh will help keep out the burrowers

Once you have laid out your mesh, then put in the layer of drainage as described above. If you have a well-drained soil around you, or under the bed then this can naturally be adapted to suit.

The drainage should then be layered over with 3-4 inches of soil mixed with compost.
The compost itself, will largely depend on what you are about to grow. For instance, if you are growing carrots or parsnips, then a light loamy compost with a good mix of sand may be required. Potatoes or leeks may require a good bed of well-rotted manure layered over the soil at the base.

It should be pointed out that the internal filling of any raised bed should not be soil alone, as this has a tendency to go rather solid after a short time. Instead mix some quality topsoil together with a good loamy compost, with plenty of organic material to keep it well drained. Again depending on your choice of vegetable a slow release fertilizer or well-rotted manure should be mixed through.

Hard work?

Ok, to be fair you may well be thinking that this sounds like a lot of hard graft – for something that is supposed to be an easy gardening method ? Well yes you could be right…However, once this part is done then you can relax and actually enjoy the next part, which is planting your vegetables. From now on it's easy street !

At this stage, and throughout the growing season you will discover just why raised bed gardening is so much easier than the traditional vegetable garden. Look over at your neighbour breaking his back hoeing between the vegetables, or digging his way through stony ground. Whilst all the time you are sitting on the side of your raised bed easily pulling out a few weeds, and plucking your ripe tomatoes.

Modification1 – Cold frame

One of the simplest and yet most rewarding conversions you can make to your raised bed garden, is to turn it into a cold-frame. This will enable you to get an early start in the growing season with all your early plants such as tomatoes.

If you stay in colder northern climes, then this can become an almost permanent answer to a greenhouse, enabling you to grow things like cucumbers, marrows, chillies and tomato plants – to name just a few.

To do this is quite simple and will any take a short time. Start before you infill the raised bed with any compost material as it will be much simpler. Material needed for the job:

1 ½ " plastic pipe
Galvanized pipe straps
½" flexible pvc tubing or similar
Polythene sheeting

Cut out eight lengths of 1 ½" plastic pipe, the kind used for domestic plumbing is just fine, measure so that they do not protrude above the wall of the bed when placed vertically inside. Next space evenly along the inside wall of the bed, pointing upward, and secure in place using galvanized pipe straps, top and bottom.
When this is done then cut the ½ " tubing to about twice the width of the raised bed. Bend the pipe and slip into place.

To stop the bent tubing slipping all the way down the pipe, simply fill with some fine gravel or even a sand and cement mix up to about 4" (100mm) from the top of the pipe. Once this is done you will have an effect similar to the covered wagons you see in the old western movies !

Next you simply fit the polythene over the framework. This is easiest done by securing the polythene sheet along one side of the frame by a length of 2 x 1 for instance. Then you simply pull over the frame when you want it covered.
The ends are a bit more awkward, but if you leave plenty material to work with, then you can simply weigh down by placing a plank on top of the polythene, and weigh down with bricks or equivalent.

The following picture below shows a smaller version of that described, but using just a part of the raised bed area.

One thing you may note is that this model can be improved by the addition of slits or vent holes in the polythene. Better still fit it with polythene that has holes in it especially for the purpose, as in the example above where only part of the raised bed is used for bringing on the seedlings.

This style of perforated polythene will prevent your 'polytunnel' from overheating.
Failing that then you must remember to remove or fold back the sheeting when appropriate. This is especially true if your intention is just to use it for the hardening of young plants.

Modification 2 – Insect/Bird netting

Another good thing about the raised bed is just how easy it can be covered up to stop the predations of birds or insects. The simplest way to do this is to follow the previous example for creating the frame effect, but replacing the polythene with a fine bird mesh.

This can be easily clipped into place in a few minutes, keeping your plants free from not only birds such as wood pigeons, blackbirds etc; but also stopping the cabbage butterfly for instance, from laying it's eggs on your plant leaves.

Another way however to do this is to follow the example in the picture.

This is an old kids swing frame, that handily fits right over the raised bed.

This in turn gives you a structure that you are able to walk around in, whilst tending your vegetables.

It would only take a simple modification to convert this into a temporary greenhouse, if you used polythene instead of nylon mesh.

Yet another way to cover up, is to raise up a post on each corner of the raised bed, link together with a 2 x 2 along the top, between the four corners.
This will give you an effect like a four poster bed, which you can then cover with your material of choice.

As you can see, there are several ways to cover your raised bed, either to use as a cold-frame or to simply protect against birds and insects. All these ways will only take a very short time, and will reap great rewards.

Modification 3 – Automatic Irrigation

Another addition to your raised bed can be an irrigation system. This can be an automatic irrigation system, or it can simply be a system that is put in place, and watered when you choose to do so. There are many watering systems on the market, but here is a simple model to follow that will do the job fine.

Place ½" polythene pipe under the soil just an inch or so, shaped like a tuning fork, with the single end at the top attached to a fitting such as a 'hoselock' click fit type. This can simply and easily be put together with 2 push fit elbows and 1 tee piece. 2 end caps close of the end of the pipes
The pipe should be perforated along its length with small drip emitters fitted every 12 inches.

After fitting a stop valve at the raised bed end, the whole system should them be fitted to a water tank - the bottom of the tank raised above the top level of the bed.

This tank can be attached to the mains water supply if needed, or it can be filled manually. with a float valve to close it of when the supply is not needed.

By turning the water valves on, the drip emitters will release a small amount of water over any given period. After some 'fine tuning' this is a system that will take away a lot of the labour attached to watering a vegetable garden.

Working Tips For A Raised Bed

When is a cucumber like a strawberry?
When one is in a pickle and the other is in a jam.

When it comes to tending a raised bed garden, there are a few differences, or subtleties compared to tending a traditional vegetable patch.

Here are a few tips for making life even easier !

Cut a strong 'spanner board' i.e. a board that is slightly longer than the width of your bed. Place across the bed, resting on each side edge. This can be used for placing your small garden tools on when working the bed, without leaving them in the wet soil. A good place also to put a glass of something cool !

Avoid standing in the raised bed. This is to prevent the soil becoming compacted, and also to prevent any chance that you will push out the sides of the raised bed, by compressing the soil over time.

If you have multiple raised beds, then put down a weed restricting fabric between them and cover with 2-3 inches of mulching material such as chipped bark. This saves a lot of laborious weeding between the beds, and leaves more time for what really matters in life.

Plants can be grown a little closer together in a raised bed, because of the concentrated nature of the feeding system.

Another good tip and one that will keep the slugs and snails at bay, is to place a copper slug tape or strip around the timber structure. Slugs hate copper because of the way it reacts to the slug mucus, so they will not cross it. If you have

no copper tape and have an immediate slug problem, try spraying a concentrated salt solution around the outside base of the bed. This can be quite effective, but do not let any into your plant bed as it will most likely kill your vegetables !

Working a raised bed garden as can be seen here is slightly different for the 'normal' way of gardening – but not so different that you need a new set of rule books so to speak. When speaking to raised bed gardeners, you will probably find that the biggest difference is the fact that they are not suffering from constant backache!

The raised bed is much easier when it comes to weed control – mainly because you have started by using virgin soil that is weed free to begin with. However even after it has been up a while, it is still much easier to weed owing to the softer loamier make-up of the soil or compost.

Even the feeding of the plants is more successful, as all the nutrients are going to the plants and not seeping away into the soil, as is normally the case.

Building multiple raised beds, if you have the space, is ideal. This allows for a good rotation of the different crops and guarantees a great harvest year after year.

Vegetables For A Raised Bed

This cabbage, these carrots, these potatoes, these onions ... will soon become me. Such a tasty fact!
- Mike Garofalo

What vegetables would I recommend to grow in a raised bed ? Just about anything !

Seriously, there is nothing I can think of that will not grow as well if not better in a raised bed, than it would in a traditional vegetable plot. The mere fact that the vegetables are raised up away from the creeping things of the soil, means that they have better protection against insects, and are not so prone to fungal disease as they have better air circulation in general.

For instance, cucumbers will grow better as they can be trailed over the edge of the bed, keeping them of the ground. Carrots are a crop where it is better to keep them raised anyway, in order to help protect from the carrot fly.
Potatoes can be easier to dig up from a raised bed, as can parsnips and any deep rooted vegetables, simply because the soil is looser.

It is much easier on the back when tending strawberries or any low growing fruit or vegetable. However, tomatoes grow exceedingly well in raised bed situations, as they can greatly benefit from a concentrated feeding regime of the type a raised vegetable garden offers.

It is a simple matter to build any kind of trellis work on a raised bed – particularly the timber models, as it is easier to secure any fixings into. Growing beans or peas in a raised bed complete with trellis or framework is a simple matter when working a raised bed vegetable garden.

Summary

As you may have guessed by now, I am quite a fan of raised bed gardening. Yes it is true that more preparation is involved at the beginning of the project, if you are going to build a raised bed. However the rewards in my view, are well worth the effort, as the raised beds that you build should give you many years' worth of service.

Another advantage of the raised bed that I have not covered here, is simply the fact that you do not need the same range of expensive garden machinery. Rotavators for instance, usually needed to dig over the soil, are not needed for a raised bed. Most of the digging work is done with the help of a small garden fork, as the soil is generally light and loamy.

In fact almost all the tools you need are simple hand tools, for light digging and pruning of your plants.

I have been asked in the past, just what is the difference between a raised bed and a planter – the answer is simple. A raised bed does not have a timber base, and therefore cannot be moved around. Planters do have a slatted base and are generally smaller, to enable positioning. Planters are generally chosen for ornamental purposes.

There are areas however where there is just a fine line between one and the other – and that is fine.

Final note of caution: Do not build a raised bed on a decking area – it's far too heavy, and will rot your deck.
Choose a smaller planter instead.

MY NOTES / TO-DO PAGE

MY NOTES / TO-DO PAGE

MY NOTES / TO-DO PAGE

Book 3

Raised Bed Gardening Planting Guide (2) - Making The Most Of A Raised Bed Garden For Growing Vegetables

BY
James Paris

Introducing The Raised Bed

What is a raised bed

If you are unfamiliar with the concept of a raised bed garden, then let me assure you that it is simplicity itself. The basic idea is that you have a (usually) wooden construction that is approximately 6 foot by 3 foot, and around 6 – 18 inches high. These are not precise measurements as this is a very flexible way of gardening and can be adapted to suit the individual's needs.

This structure is then filled with a compost material, made up of part compost and part soil, mixed with plenty of good organic matter throughout. This is of course adaptable to the type of vegetables that you intend to grow – more on that later on. Never just fill the raised bed with soil only, as this has a tendency to go very compact, making it difficult to work, and restricting general growth of the plants.

Why a raised bed?

There are many advantages to growing vegetables in a raised bed as you will see from the following list:

Easier to work and maintain than a traditional vegetable garden.

You can grow the same selection of vegetables that you can in a garden, but with only using a fraction of the space.

Far better height to work with, especially if you suffer from back trouble or are otherwise physically impaired.

No trouble with weeds, meaning that more time can be spent tending to the vegetable crop itself.

Easier to keep free from garden pests including insects, and vermin such as rabbits and mice etc.

Great for specializing in just a single crop, or using the 'square foot' method of growing many types of veg in a very limited space.

Can be modified to make a cold frame, or netted over to protect vegetables with minimum of fuss.

The growing season for a raised bed garden is longer as the soil warms up quicker than the ground soil.

You can even build a raised bed garden on a hillside, making one side higher than the other, thus compensating for the slope and giving you a flat growing surface.

A raised bed does not get all compacted with people walking on, therefore no need to double-dig to loosen up the soil.

***

Some publications state that it is not feasible to grow tall plants that need support in a raised bed garden owing to the fact that the soil is not compact enough to hold the canes or structure in place, This is however nonsense! As will be established further on in this work, virtually anything that can be grown in a traditional garden, can in fact be grown (usually a lot more productively) in a raised bed garden.

Building A Basic Raised Bed

To build a raised bed need not take a lot of time or indeed money, as the fundamentals are fairly simple as well as the materials being quite diverse. However for the sake of the uninitiated! Let's look at a basic 6' x 3' x 12inch raised bed structure, and how exactly you would go about building it. Simply go to your local timber supplier and purchase..

6 lengths of 6 inch x 1 inch timbers, at 6 foot long.

6 lengths of 6 inch x 1 inch timbers at 3 foot long.

6 lengths of 3 inch x 1 inch timbers at 12" long.

Actual timber decking material can be excellent for the planking material needed..

On top of this you will need to purchase or otherwise source the compost material to fill your raised bed. As mentioned earlier, never fill the raised bed with soil only, as this will have a tendency to compact together.

Mix soil with plenty of organic material to make the best infill material. Obviously this material will change slightly according to the vegetables you intend to grow. For instance if you are growing root vegetables like carrots or parsnips, then you would choose a very loamy soil with a good sand mix through it.

When you have cleared an area of ground for your raised bed structure, then simply mark out the 6 x 4 area and level it off. Note: No need to dig up or turf an area before laying down your raised bed frame; simply cover the turf or weeds with several layers of newspaper, before adding a bed of gravel for drainage and then the compost mix. To construct the timbers simply nail or screw the timber planks to the 3 x 2 timbers. Two planks for each side and each end. The long sides will have a 3 x 2 inch timber in the center as support.

The finished result will be a 6 x 3 foot timber frame, at 12 inches in height.
There is no need to over complicate this technique, contrary to some opinion. All you are doing is making a 12 inch high timber box with an open top and bottom, so that you can fill with compost and grow vegetables – end of story !

There are of course other ways to do it, and you may see some other examples and a much longer-winded explanation in my other kindle book 'Raised bed gardening – growing the easy way'; as well as a range of materials that can be used to make your raised bed.

Here is an example of a raised bed made from simple heavy timbers, just secured at the corners:

Planning Your Bed

Think ahead:

Like all things in life it pays to think ahead before diving rashly into something, this is the same with gardening.
I mentioned in the previous chapter that the type of infill for the bed, would depend on just what exactly you intend to grow there.

Of course many vegetable types will grow in the same soil, and so it is not a real problem. All vegetables need plenty of organic material included. This gives the plants the much needed nutrients, plus it provides a good medium for the roots to grow through and expand.

If you are into measuring the pH levels, then they should fall somewhere between 6-7 pH for ideal growing conditions.

Raised bed trellis

Another reason to think ahead however is that if you are thinking of growing climbing plants like peas or runner beans, even cucumbers; then you will need to construct a growing frame to support them. This is fairly easily done with a raised bed garden as the above picture demonstrates fairly well I think.

This framework is made out of 1 inch plastic plumber's pipe that is fairly easily obtainable from most hardware stores or plumber's suppliers.
The general idea is to do this before adding the soil and is done in the following way.

Simply fix short vertical lengths of 1 ½ inch pipe along the inside of the frame about two feet apart, with some galvanized strapping. Into this slide down your 1 inch plastic pipes and connect with T junctions as per the picture.
As you can see here this has been extended in the centre and gives the perfect frame for growing any number of different vegetables that require a frame or trellis work to climb on.

Connect together with nylon cord to give more support where needed, this will support your heavier plants and strengthen the overall structure.

If you leave this till later on then you will have to either trust in your light loamy soil to hold the piping in place (bad move) or you will have to empty at least some of the soil out at each upright in order to fix in your short vertical upright pipes that are the holders for the frame.

This is an idea that is better suited to the higher frame heights to be fair, but can be almost as effective in the shorter 12 inch high versions.

Raised Bed Hot House

By following the previous chapter and building your trellis work, you can also make a very effective hot house or indeed cold-frame, depending on your usage of the thing!
To do this, just build the frame as before but exclude the upright part of the framework.

This makes it very simple to cover the structure with polythene that can be fixed to one side of the structure permanently and have it secured to a 3 x 1 inch strap along the length of the other side. This allows you to lift the polythene up and over, to gain access along the whole length of the frame.

Alternatively you can fix the sides and just leave the ends folded up in such a way as to allow access through the ends only; though this is probably only practical with the shorter 4 foot long models as access from the end to the centre is possible with a little stretch from either end.

Bird protection

Again this model can be used very effectively for bird protection, by simply covering with a fine nylon mesh. This can easily be fixed in such a way that it can be removed for access. A light fleece can also be applied over the frame to protect against frost or indeed to shade from the sun, under certain harsh conditions.

Another alternative is just to build a simple wooden frame onto or around the raised bed itself. This can then be covered very effectively for bird protection.

Six Foot - square foot

Another interesting idea for a raised bed garden is to follow the principle of the 'square foot' gardening method. Square foot gardening is simply taking a structure of four foot by four foot, and separating them into one foot squares by

means of a simple framework placed over the top of the area. This can also be done with taught nylon string as in the above example.

This gives sixteen potential 'mini plots' to work with. The idea here is that a family of four can actually produce enough vegetables throughout the growing season to feed them all comfortably and cheaply.

Unbelievable as this may sound, it is indeed possible if enough thought has gone into the preparation and a good rotational plan is followed. One of the good things about this plan (and there are many) is that the plot should never need to be artificially fertilized as the vegetables in each plot take only the nutrients that they need, and as they move around they leave the other nutrients for the plants that come along behind.

As an example of this, you have grown beans and peas. They take nitrogen from the air, and leave it in the soil. Thus it is good to let these plants die at the end of their growing season and in turn fertilize the soil with a nitrogen rich environment for vegetables such as Cabbage Cauliflower or kale that love this environment.

As in fact do potatoes, though they should not be planted alongside brassicas as they prefer different pH levels.
This is in fact the traditional crop rotation method in miniature and works very well for the four foot square garden.

If we simply take this method and place it into a six by four foot raised bed, then you would have twenty four potential planting areas. This is more than enough for the average family needs of vegetables, if it is properly handled.

Planting Out Your Raised Bed

What to plant

When deciding what to plant in your raised bed, there are a number of things to take into consideration. These could include the following:

How many raised beds do you have to plant? More than one bed means that you can have a larger crop of one single vegetable kind.

How big is your family, or how many do you intend to feed?

Does your raised bed get full sun, i.e. at least 6-8 hours per day?

What do you like to eat? This may sound like a silly question; however it is easy to grow something just because you can. Only to find that it is wasted at the end of the season because you do not really like Brussel sprouts for instance!

Did you use this bed last year, and if you did, what did you grow on it? Good crop rotation is key to getting a great harvest.

As you may understand, what to plant is a question that has many answers depending on your individual wants and needs – as well as the wants and needs of your friends and neighbours!
For this reason I will cover a few different planting regimes that hopefully may give you some ideas of your own.

Crop types

First of all we must divide the crops into their respective families, in order to get the best out of the soil conditions they are planted in.

Root Crops: Potatoes, carrots, parsnips, beetroot, fennel, celery

Brassicas: Cabbage, Brussel sprouts, cauliflower, broccoli, radish, swede turnip

Legumes: Peas, mange tout, French and broad beans

Alliums (onion family): Shallots, onions, garlic, chives, leeks

Solanaceae: Aubergine, potato, tomato, peppers, eggplant

Cucurbits: Cucumber, squash, pumpkin, melon, marrow

Miscellaneous: All fruits, lettuce, herbs, sweetcorn, chicory, asparagus

Though this is by no means an extensive list, it does give a good selection of the most commonly grown vegetables, and is more than enough to get started with!

Watch your height
One of the things that is easy to over-look, but most important, is to be aware of the height of the plants. Plant the high plants or climbers in the north end of the bed, that way they will not shade the rest of your crop from the all- important – life giving – sunlight.

In other words, if you have a raised bed that is broadside on to the sun for most of the day, then perhaps a frame built along the back of the bed would be a good idea. This would enable you to grow runner beans, peas or cucumber plants, making an excellent backdrop to the vegetables in the rest of the bed.

Peas would have the added advantage of adding nitrogen to the soil that would ensure a great harvest from the likes of cabbage or cauliflower that are particularly nitrogen hungry plants. It is also true however, that all plants like some nitrogen in their diet of nutrients, and so this technique would benefit just about everything.

A single crop

Planting a single crop in a raised bed is usually only done if you have more than one bed in which to grow your vegetables. After all what is the point in growing just one crop, unless of course you are a fanatical pumpkin grower, or just cannot see past a good cucumber harvest!

If however you do just have one crop in mind then that is fine. Your choice however should not just be on what you like, but rather on what you have grown in the bed previously – if at all.

Good crop rotation is not only important as to the nutrient value, but also for pest control and issues such as blight and fungal growth. If you have suffered a case of potato blight the previous year for instance, then you would not want to try and grow the same crop again this year as the potato blight virus can remain in the soil for several years.

Similarly it is not a good idea to grow onions in the same bed for more than two or three years running, if you want to get the most out of your harvest.

Even with a single crop as your main crop however, you can still grow a 'companion' crop if you mix them correctly.

Companion crops are vegetables that get on well with other vegetables such as onions, carrots and lettuce or spinach onions and brassicas, being that their nutritional needs are similar, but their root systems collect the nutrients at different levels and so are not truly in direct competition with each other.

As well as giving you a variety of vegetables, this is a good idea for things such as pest control and less weeding as the veg blocks out the light to the weeds; and different crops attract different insects thereby helping control the spread of the insects themselves.

Antagonist crops are crops that do not do so well together such as Alliums (onions and garlic) with peas and bean crops. Crops such as beetroot get along with most plants and so can be planted quite successfully between onions or leeks without any problem.

Mixed crop examples

If you have just one raised bed and are trying out several varieties of vegetables to give you a good mixed crop, then some planting tips could include the following regime:
Tall plants to the back, on a fixed support of some kind where needed.

This could be planted with tomatoes, or peas, runner beans or cucumbers if the climate allows.
Sweetcorn could also be used here and companioned with lettuce that will take advantage of the shade from the corn.
To the front of the raised bed could go any low-lying crop such as carrot, beetroot, parsnips etc.

If peppers are your main crop then you could grow spinach between the pepper plants. Again similar to the lettuce and corn example, the spinach will take advantage and flourish in the shade provided by the peppers.

If you stay in a cooler climate such as the UK, then you could try covering a portion of your raised bed with a frame similar to that pictured earlier, and cover with polythene. This will enable you to grow tomatoes and cucumber plants for instance, maybe even some sweet peppers – so expensive to buy in the supermarket!

To get the most out of a mixed crop the idea is about looking at the different vegetables needs regarding hours of sunlight and nutrient requirements. A good idea is to plant vegetables together that have different root systems, as mentioned earlier.

Planting deep rooted vegetables such as carrots or parsnips, means that you can companion them with shallow rooted vegetables such as beetroot, lettuce and arugula (rocket) to make for a good health selection.

Ultimate mixed crop

We can hardly discuss the idea of a mixed crop, without including a piece on the square foot example of growing mixed vegetables.

As mentioned, this is a system that is intensive farming gone to the extreme – but in a good way! The idea is that the old method of growing vegetables in a garden plot, where everything is planted in rows, is made redundant.

Instead we have a situation where a plot of ground – raised or otherwise – is set out in squares 1 foot square. Within these mini-plots a range of vegetables are planted, according to your needs and other deciding factors such as weather, nutrients etc.

The 'sales pitch' if you like, for this style of gardening include the following points:

A large percentage of a traditional vegetable plot is wasted owing to the fact that you must have pathways between the vegetables for weeding, harvesting etc.

With a proper rotational method of planting, you need never use fertilizer again, as the crops will feed themselves.

Far less waste than a traditional garden where all the cabbages, or cauliflower or whatever, is ready at more or less the same time, and has to be frozen, canned or given away.

Similar to a raised bed, the square foot method does not need such intensive care, due to the close proximity of the plants and the ease of gardening methods, particularly if it is used with conjunction with a raised bed garden.

One of the main strengths of this gardening method is the fact that it uses up very little space, especially if you go for the four foot square method. This means that despite this small space you can grow a complete range vegetables and have them all maturing at different times, depending of course on the length of the growing season.

That said, you have to keep a keen eye on any disease or insect infestation, potentially caused by the close proximity and therefore poor air circulation, in this type of growing environment.

A square foot garden design lends itself very well to cultivating a full spectrum of herbs available. This in turn will ensure a garnish for every meal you prepare as well as some great tomato and basil salads for example.

Pest control

One of the things that mixed planting is good for is pest control, as we have mentioned the combination of different plants can confuse the pests and help control their breeding program. However there are certain plants that you do not want to mix together because they encourage the same pest. This could lead to an exacerbation of your pest problems, instead of the opposite.

For instance, if you consider pest control to be a major objective in your growing regime, it is not wise to grow sweetcorn and tomatoes together as they both attract the corn ear-worm, also known as the tomato fruit worm. Tomatoes, eggplant, peppers and potatoes are a favourite of the Colorado beetle; while squash, melons and cucumbers are a favourite snack for the pickleworm.

Pest control in a raised bed

Controlling slugs in a raised bed situation is particularly easy, compared to the traditional growing methods. Simply fix some copper tape around the edge of the raised bed; slugs will not cross copper as it has a chemical reaction to their slime. Copper paint will do as well.

There are of course the usual tips about beer traps or bran and vinegar to attract them. Either method works by setting a trap including these ingredients and hiding it under a roof tile or similar. The slugs are attracted to the traps where they can then be disposed of.
Carrot fly prevention is one of the bonuses of a raised bed as carrot fly do not tend to go over 18 inches or so, as a general rule. This means that the higher raised beds have a good advantage over the battle with carrot fly. One of the top tips to keep away the carrot fly include extending your raised bed frame

height by about one foot or so, then adding some fine insect mesh around it.

This can be very effective against the carrot fly, as they are low flyers as mentioned, and so this will prevent them from dropping by and laying their eggs on your young carrots.

In general however the advantage of the raised bed, is that it is fairly easy, especially if you have developed the cold-frame idea, to flip a fine nylon mesh over your crop. This will prevent the egg-laying butterflies and moths from getting access.

It will also prevent the birds from helping themselves either to the young seedling or indeed your strawberry harvest.

Rodents:
To keep out burrowing rodents such as moles or gophers, then you need to lay down some wire mesh on the bottom of the raised bed before you infill with the compost material. This should be a fairly sturdy galvanized mesh with about 1inch holes preferably.

This will prevent most critters from gaining access to the bed from below.
The smaller rodents such as mice or voles have to be kept away with the usual fine mesh or even fleece material, if you want to prevent them from nibbling on your young peas and beans.

I have had some success with a battery operated sonic device for chasing mice. If you just place one of these devices anywhere in the bed, cover against the weather, and turn it on. Results have been mixed according to some of the comments I have had from others but it may well be worth a try at least if you have a bad mouse problem.

Summary

As you may have guessed by now, I am a big fan of raised bed gardening for all the reason covered in this and other publications. It is difficult to find a downside to this type of gardening, unless perhaps you want to include the work and expense of setting up a raised bed in the first place.

However even with that said, gardening does cost some dollars to set up. Nothing in this life is free, and everything you commit yourself to do usually has a cost in time, material or just plain mental thought!

The bottom line though is that gardening, whether in a raised bed situation, or just a traditional back yard situation; has many things to commend it. From growing flowers to beautify your life and the life's of others, to growing vegetables to feed your family and perhaps the rest of the neighbourhood, gardening is good for the soul and the stomach!

Many schemes have been set up now to encourage raised bed gardening in particular, as a way to bring whole communities together. Growing and sharing the work of your hands has a real long lasting value in any community, and is a great way of getting to know your neighbours. Children in particular seem to have a real affinity to working in the garden, especially if you teach them from an early age – before the teen spots begin!
This can be a great way to build up in your children a real respect for nature and the way things are in reality, not as they are on TV or the supermarket shelves.

MY NOTES / TO-DO PAGE

MY NOTES / TO-DO PAGE

MY NOTES / TO-DO PAGE

Book 4

Vegetable Container Gardening – Growing Vegetables In Containers & Planters

By

James Paris

Chapter 1: Planting – Where Am I Coming From ?

"no-one is born in a vacuum and we are all molded by the circumstances in which we find ourselves" ***anon***

I was born and brought up in the kind of family where money was always on the scarce side, and I was no stranger to making ends meet with whatever came to hand. Consequently although I may be accused of being on the mean side with the old cash flow, I would say in my defense that I was really just very careful with the somewhat limited funds at my command.

The benefit of being raised under these somewhat restricted circumstances, is simply that you are soaked deep down with the understanding that money is very hard to acquire, and painfully simple to throw away.

What does that have to do with vegetable container gardening you might ask ? Simple really; planting vegetables in containers can be a great way to get maximum value out of a limited space or resource, and if it is utilized properly you can even make junk turn you out a handsome profit.

Some people are put of the idea of growing their own vegetables because they do not have a garden, or a plot of land big enough to raise vegetables on. This is however a misapprehension, as the fact is that by following the right guidance you can provide a good supply of vegetables from a very limited space by utilizing the ideas in this manual.

As a child of about 11 years old, my father decided to move us lock, stock, and barrel to what amounted to a smallholding just a mile or so up the road from where we stayed in a rented council house. For us as kids it was a mixed blessing because on the upside we now stayed in a large house, with loads of garden space and some old out-buildings to play in.

However the downside was that from the get-go my father decided that we would learn his ways, from his childhood. This was not good news to us, as with this move he decided that we should 'pull our weight' and get to work on the farm.

Now you might think that with plenty of land available to plant in, we would have no need of growing vegetables in containers. However this was not the case as the fact is with a family of nine children to feed, every space has to be utilized to its best advantage, and every money-saving technique put into place. My parents were not wealthy, and so with good Scottish prudence put us to work around the smallholding.

Amongst the things I hope you will find interesting in this book, apart from tips on planting in containers; are things like how to renovate an old greenhouse on a tiny budget, how to grow vegetables in planters and raised garden beds as well as produce tomatoes, cucumbers and even grapes from an old grapevine!

Life "down on the farm" was not a barrel of laughs, it has to be said. There were times when my brothers and I would have done anything to get out playing football with our friends, or get away causing some mischief somewhere.

Instead however we were stuck on the farm, hoeing weeds out of the vegetable plots or mucking out pigs and horses. Not to mention chickens, geese, ducks, rabbits and all the other things associated with running a smallholding.

Not long after we moved into this smallholding, which was basically an old farm house with a couple of acres of land attached, my father decided that he was going to use a part of this ground to break up cars, as he was actually a car dealer by this point.

You probably think that I am wandering away from the point right about now ! However it is all relevant, as will become clear later on in the book.

We soon discovered that even scrap cars can have their uses around the vegetable garden – strange but true. They have their dangers also when young children are around, as we soon discovered to our cost – however that is another story !

General repairs around the place had to be done of course, and this was when it was really discovered that I was particularly good with my hands (The brain was on vacation most of the time, so it was as well I was good at something!) Needless to say that rather than pay for experienced men to handle the tasks around the place, I was put into the "front line of defense" when a shed needed repaired or built.

As you can imagine on an old farm, there is always something needed repaired, so I was kept busy most of the time, much to my annoyance.

Yes of course mistakes were made – I was only 11 years old don't forget – but throughout this "school of hard knocks" experience, I learned a lot.

How to grow vegetables in containers – I first of all had to learn how to build planting containers, on a very limited budget, as you may well imagine.

Chapter 2: Time Rolls On

"Time is the coin of your life. It is the only coin you have, and only you can determine how it will be spent. Be careful lest you let other people spend it for you." Carl Sandburg

Yes, the years have moved on at an alarming rate, and now at 54 years of age I find myself writing this ebook, attempting to put some 43 years of knowledge into a few thousand words !

Most of it is thankfully irrelevant to this book, so at least you may be spared the ramblings and moaning's that are the lot of most people (so I am told) of my age.

Getting back to the point of this exercise; growing vegetables in containers or planters, is not rocket science. No matter what some of the manuals tell you, growing or planting anything, let alone vegetables, just needs the will to do it and the right conditions to do it in, then the results will come naturally.

Granted, if you are intent on growing prize tomatoes, or huge award-winning marrows, then a certain expertise is involved. Being best at anything does not usually happen by chance, but rather by real commitment and dedication to the task – no matter what it is.

During most my so-called childhood, along with the rest of my brothers, I was busy between school and the farm. The summer was a curse of sorts to us, because while our friends were 'out and about' we had to work in the lighter evenings. At least during the darker winter evenings we could be out with our friends.

However, years later it turned out that I had in fact learned a few things about self-sufficiency. How to make the most of very little, and how to survive through tough economic times – like now.

Learning how to grow vegetables, was just a part of my upbringing. However it was a part that has an increasingly significant place in the world in which I find myself today; and one which I hope to share in this book.

Scrap Cars & Recycling

I mentioned earlier, that even scrap cars have a role to play in the world of container gardening. Well the fact is that one of the most difficult parts of a car to re-cycle, are the tires. The steel wheels were always easy enough, as they are indeed scrap and so they were worth collecting and taking to the scrap merchants.

The tires though were another problem. Now however that problem can turn into an advantage, at least in a small part, because the tires can very easily be made into very good and productive planters, at zero cost to you.

Simply go down to your local car-breakers and most of the time they will gladly give you as many tires as you need, as they have to pay to get the tires taken away and disposed of.

When choosing your car tires however, be sure that they are all the same size, obvious I know – but it's very easy to forget the obvious sometimes and be left with a problem you don't want.

The different kinds of containers that can be used for container planting, as well as the construction methods employed where needed; will be discussed in the next chapter, as will the soil or compost mix and numerous details that must be taken into account to get the most out of your container garden.

Chapter 3:Planter Types and Ideas

Necessity, who is the mother of invention. ***Plato***

Often times the best ideas simply come when you have to think "outside the box" usually because you just cannot afford to pay shop prices for what you want. Such is the case for our first planter, discussed in the previous chapter – the tire planter. This must be one of the simplest planters to construct as you are really just stacking tires one on top of the other.

As mentioned the tires should all be the same size, unless you want to have a tire that looks like it has a pot belly ! Normally there is no need to link them, as the weight of the soil itself will usually keep it all together.

However if you want to play it safe, all you need do is lay the tires one on top of the other until you have achieved your desired height. Then link them by boring through them with a drill or sharp awl, in three places. Tie them together with galvanized tie-wire or nylon string – job done.

The height of the planter depends on whether or not you need deep compost, this really depends on what you intend to grow.

A word of caution. Make sure that there is no contamination of the tires when you collect them from the breakers yard. Diesel in particular will kill your plants stone dead, ruin your entire day really!

Tire planter for Potatoes

Tire planters make particularly good potato barrels. If you have not heard of this concept in potato growing, then it is simply this. Growing potatoes in a small area, is quite impossible using traditional methods, and this is where the potato barrel comes in.

By using your tires, start by placing three tires in the area where you plan to grow the potatoes, one on top of the other. (it cannot be moved later). **Do not link** these tires together, you will see later why.

Make sure you have some drainage at the base (broken clay post or gravel will do fine) then fill up to the height of the second tire, with a good compost material. Place your seed potatoes in place with shoots facing upwards (usually two or three seed potatoes will suffice) and cover over with a further 2-3 inches of soil or compost.

Place another tire on top, and wait for the shoots to break the surface by a further two or three inches. Put in more compost up to the top level of the shoots, with just a little breaking the surface.

Repeat this procedure, till the tires are about three feet tall, or starting to get unstable. When your crop is ready to harvest, you can simply pull apart your planter, easily revealing the mature potatoes ready for harvest.

Now you see why we did not tie the tires together, it would be a bit of a nightmare to harvest the potatoes if we did, as the nature of the tires means the compost gets a good grip inside, being filled with potatoes. This can make it incredibly difficult to get them out of the planter.

Other planter types that suit the potato barrel model are, the **box wood planter** that can have the sides removed to reveal the crop when ready. Then there is the **simple plastic bag** method. This gets cut up the side to easily remove the potato harvest.

Then there is the **barrel method** where to harvest the crop you have to tip over the barrel completely to get the potatoes out.

The Raised Brick Potato planter as in the example below, is especially effective and easy to harvest if you lay the bricks dry like a dry-stone wall.

Just remove the bricks to reveal your potato crop when the time is right.

You can read all about the finished results of this particular experiment on my blog post at http://planterspost.com/potato-planter-results/

All of these planters follow the same idea as to raising the soil level as the potato grows. All very successful, and extremely satisfying when it comes to harvest time.

Traditional Timber Planters

Timber vegetable planters are very easy to construct, and indeed can be reasonably inexpensive. One of the cheapest ways is to go to the local sawmill if you have one, and as if they have any 'barks' available. These are the offcuts that are trimmed of the tree before it is cut into regular shaped planks.

I have built many things in the past for virtually nothing by using barks instead of expensive cut planks. Not just planters, but garden sheds can be built using this method.

Building with barks can look great, and really fit in with the surroundings a lot more naturally than finished timber can.

To be fair, there is some extra work sometimes, just squaring them of a bit. Depending on the load there can be a fair bit of waste also. However for the barks that are no use for construction, simply cut them to size for use in the log burner !

Building with sawn timber however does have it's advantages in that it lasts longer and is quicker to construct, just because the timber is all of a uniform size.

Timber planters come in all sizes, but the construction methods are basically the same. It is not rocket science, and is in effect four or six posts, with planks nailed up the sides to the desired height.

A basic timber planter for instance may be six feet long (1,500mm) by 18 inches (450mm) wide by one foot (250mm) high. It is advisable to re-enforce with an upright post ever three feet, and so this planter for instance would have 6 posts, one in each corner and one in the middle on each side.

The base of the planter is best made from plant friendly treated timber **(not creosote, which will kill the plants.)**, about 1 inch (25mm) thick at least. This will ensure a good few years from the planter.

The base should be raised a couple of inches from the ground, and have a few holes bored for drainage also, other-wise your plants will become waterlogged, which is no good unless you are growing rice !

Top tip – do not use plywood or chipboard for your planter as it will just burst. Stick to sawn solid timber if you want it to last.

A vegetable planter of this kind is really another adaptation of a raised bed, with the only real difference being that a planter can be moved around (though it will be extraordinary heavy when full), and has a raised base built in. A raised bed garden is fixed in place as the posts are sunk into the ground, and the raised bed is generally a wider construction.

Using an old wheelbarrow ?

Old wheelbarrows make excellent planters, especially if they still have a working wheel on them, as this enables you to follow the sun (or the shade), and makes for a truly mobile planter. A great advantage, if for instance the sun is only a short time in one area of your garden.

Be sure you punch a few holes in the base for drainage, as they easily get waterlogged. Painted up, a wheelbarrow can look good as well as "produce the goods" when it comes to growing vegetables or plants.

Another idea for a wheelbarrow, particularly if the base is rotten, is to prop it up on end against a wall, handles in the soil. With this you can now grow a climbing vine up the wheelbarrow, using it as a makeshift trellis. This can be really effective once the plant has taken hold.

If you do not have an old wheel barrow, then why not make a wooden one ? This can be particularly effective, and a real addition particularly to an ornamental garden, where the ascetics are important.

Traditional ornamental pots.

If you want to grow vegetables on your patio are for instance, it is probable that you would like something a bit more "pleasing to the eye" than a re-cycled ten gallon paint tin, or maybe an old kiddies bath (both can make very effective planter).

This is the time where you might spend some cash and buy an ornamental plant pot for a patio. Be sure that it is glazed if it is liable to be left outside in the frost, otherwise it will burst the first winter unless you overwinter it inside.

One of the advantages of an unglazed ornamental pot is that you can age a new pot quite effectively by brushing it over with yogurt or just milk. This will encourage the growth of algae to give it an old look.

Pots can look particularly effective if you are growing climbers like peas or beans for instance. Simply make a pyramid shape out of three canes, sunk into the compost and tied at the top. This will produce a good harvest and look very effective on your patio.

A word of caution. Clay pots can dry out very quickly, so be aware of this and water accordingly.

Black plastic pail planters

Ok, so a black plastic pail does not make for a very attractive planter, granted. However the black plastic has an added advantage that it absorbs the heat from the sun, which in turn transfers to the compost.

This makes for good growing conditions for the likes of marrows, cucumbers, chilies and a host of other vegetables that prefer a bit of warmth in the soil.

Again, be aware of the fact they tend to dry out quite quickly and so need fairly constant watering.

Hanging basket planters

Most people think of hanging baskets only for flowers, however they can be very effective also for vegetables or fruit such as strawberries. The fact they are hanging up in the air means that the dreaded slugs won't get to munch away at your strawberries – a fruit which they seem particularly fond of !

Wire frame hanging baskets are very cheap to buy, and can last many years if taken inside over winter.
One tip for hanging baskets is to make the chains a little longer than usual (you can buy chain at most handyman stores), and plant beans or peas to climb up the chains. This is a great system that will produce a good harvest very cheaply.

Planting in the gutter !

Ok, this is perhaps more for raising seedlings than growing a mature crop. Nevertheless old guttering is great for raising young plants, because you can take them out of the guttering with a simple cut with your trowel. This means less disturbance for your young plants, which in turn means that they will transplant a lot more effectively, leading to a more successful crop.

There are a few things that you could grow to maturity in old guttering – lettuce for instance, or maybe a miniature herb garden ?

Guttering does tend to get waterlogged quickly when used outside, so best to run along the base and drill a few holes for drainage.

Timber window boxes

These are constructed in very much the same way as the timber planters mentioned earlier, but are usually built to fit the particular windows. Be aware when building, that window sills tend to be sloped away from the house. This means that when you attach the wooden runners on the base to keep it clear of the sill, you put a corresponding slope on the runners so that the planter will sit level.

It may seem a small point, but it is simple to implement during construction, and means that you are not having to prop up the box with bits of slate to keep it level.

Since window boxes are meant to decorate the house, in most cases. A good option for planting here is the strawberry plant as they will not block out the sun from the interior of the house, and do produce a nice flower as well as great fruit of course.

Oh, did I mention – remember to water regularly as in all the examples before !

Use The Compost Bin!

Ok, admittedly this does not look all that attractive, however the fact is that nutrient-greedy plants like zucchinis will thrive when planted on top of the compost bin – as is in the case below.

Chapter 4: Container Planting Top Tips

Gardening requires lots of water - most of it in the form of perspiration. *~**Lou Erickson***

Compost for containers

It is important when planting in a container that the conditions are just right, if you want to maximize your results.

Containers in particular, because of their very nature are more prone to leaving a plant root bound, if they have been planted to close together. Just in the same way that a pot-plant can get root-bound.

Other things to watch out for when planting in containers is watering, mentioned several times in this article already. It is nevertheless one of the easiest things to get wrong, and so suffer a poor harvest or worse – a dead one !

How to avoid over-watering as well as under-watering will be looked at in this chapter.

The compost in planters, is just slightly different as that when growing straight into your garden bed for instance. One of the main points to consider is not to use soil from the garden in a planter, as it tends to set hard in the planter.

Instead use a good mixture of compost and other material to keep the soil loose and aerated, more on this later.

What a plant needs

First and fore-most a plant needs light to thrive, yes it's true that some need more than others, but fundamentally vegetables in particular need plenty of light.

The leaves of the plant soak up the light, and are in fact the 'engine room' so to speak of the whole plant. The roots soak up the water, but the light is the real energy source that feeds the engine.

That said, this is one of the biggest advantages of container gardening. Simply that you can place your containers where the plant is getting the best light conditions available.

Bear in mind that "full sun" is considered to be around 6 or more hours per day. So all your calculations should be based around that figure.

Water conditions we have already explored, so no need to go into that further at this point. Just bear in mind that a plant will need different levels of water according to its needs, and the particular stage of growth. IE a tomato plant that if full of fruit will need more water than one that has yet to produce any. It's not rocket science !

Feeding your plants

When it comes to feeding your vegetable planters, then the debate goes on as to whether or not to go organic, and dump the chemical fertilizers altogether.

Personally on this issue, I use organic fertilizer whenever I can, but understand that not everyone has access or the time to feed their vegetables with organic fertilizer on a regular basis.

Feeding organically, with the use of well-rotted manure for instance, takes a bit more preparation than just throwing a handful of 'growmore' at your vegetables. However I believe the results are worth the effort.

Well prepared compost should include where possible, a mix of well-rotted manure or composted material from your kitchen for instance. Mix this through with your 'store bought' compost if you have none of your own. This will ensure that your vegetables get a good start.

If you are using your own compost from your composting bin, be sure to take the compost from the bottom of the pile. This is 'the good stuff' and should be crumbly and definitely not wet or smelly. This would indicate the compost was not ready for planting.

Basically what you are doing is adding a balanced diet of .N(Nitrogen), .P(Potassium) and .K (Phosphorus) – the key ingredients to a happy plant !

If you are using a chemical fertilizer then you should look out for this designation on the box, where N.P.K is usually marked according to the mix.

Nitrogen loving vegetables like cabbage or spinach for instance like a higher percentage of .N in the mix, whereas peas and beans get their nitrogen from the air and so do not need a strong nitrogen based formula.

Most marketers of chemical fertilizers nowadays have this marked on the box, even designating the vegetable types that the fertilizer is best for.

Filling your planters

As mentioned earlier, the material used to fill the planters cannot include your garden soil, unless it is of exceptional quality and of a loamy nature. The reason being is that soil in a planter tends to go firm, and makes it difficult for the plant to thrive.

To begin with make sure that whatever planter you decide on, has holes in the base for drainage. Planters do tend to get waterlogged very easily if there is no drainage built in, leading to stem rot or worse.

To complete the drainage part you have to scatter some gravel or crushed stone about one or two inches deep along the bottom of the planter, old broken clay pots are ideal for this. This will simply prevent the drainage holes from getting clogged up and made ineffective as a result.

Next is of course to add your compost mix, hopefully you have made up your own quality compost with a good mixture of compost and well-rotted manure. This will add the needed nutrients and also help keep the compost 'loose' enabling maximum root growth.

Hanging baskets as well as unfired clay and wooden planters can easily get dried out, so the addition of 'pearlite' or other water-retaining products is advisable when making up the compost mix.

If you are planting in a hanging basket then of course the gravel or broken pot idea is redundant, simply line the basket cage with coconut fiber liner or whatever equivalent is available.

Next it is advisable to line with a black plastic liner to prevent the moisture escaping entirely. Poke two or three holes to allow some drainage.

If you are using the organic method versus growmore for instance, then mix your compost depending on the level that your vegetables require. Potatoes for instance like a good mix of manure, but peas on the other hand do not.

Depending on just what vegetables you plan to grow will by and large determine the kind of 'growing compost' mixture you should use to fill your planter. For instance root vegetables like carrots prefer a more sandy free-draining mixture.

Feeding

Once you have your vegetables planted, then you must of course put in place a feeding regime if you are to get the best out of them. Because of the limited nature of the space involved in most planter's, feeding is highly recommended.

Mostly this is done with a liquid fertilizer during the stages of growth, perhaps with a light feed once per week or so, depending on the liquid feed used – refer to the manufacturer's instructions here if in doubt.

A slow feed granular mix can also be used, simply scatter lightly around the plant base and the watering in will do the rest. The main thing with the chemical feeds is not to over-do it, as this will do more harm than good.

Manure caution

Just a word of caution, is perhaps needed here. Never use dog, pig or cat manure on the garden or in your compost heaps or bins.
Certain parasites like worm larvae tend to stay for a while in this type of manure, and so it should never be used for growing vegetables.
Also it is a wise move if you are bringing in manure to know what the source is, is it a pig farm for instance ?

For the record, sheep, cattle and horses make the best manure overall.

It goes without saying, that you should also keep pets out of the garden, especially cats – but you knew this already, right !

Finally, if you are spreading manure on the surface of the garden, then be sure that it is well rotted, and never spread around fruit like strawberries or vegetables like marrows and cucumbers as they will be laying amongst it, increasing any chance of contamination.

Compost Tea

Tea Recipes

Making a 'quick brew' of compost 'tea' is an efficient way to give your plants a quick boost, especially when they are at the fruit-bearing stage.

Here are a few recipes that you can try out – extracted from my book on Square Foot Gardening.

Compost Tea: Place mature compost into a large drum, filling about half way. Fill to the top with water. Stir thoroughly then let this mix brew for a period of 5 days or so, then strain of the compost and add the liquid to the base of the plants.

Horse Manure Tea: Follow the recipe for the compost tea, but add only 1/3rd manure and two thirds water. I have found this feed particularly effective for Tomatoes.

Comfrey Tea: Rich in potash and nitrogen, Comfrey is worth growing in any patch of ground for this ability alone. Add a large bunch of chopped-up comfrey to your water bin, place a brick on top and fill with water. Let it brew for about two weeks before adding to your veggies.

Nettle Tea: This tea does not contain much in the way of phosphates, but has usable amounts of nitrogen, iron, and magnesium. After donning heavy gloves to avoid the stinging nettles! Choose young plants without seeds or roots and put a large clump into a pail. Chop up with sheers and ¾ fill the pail with water . Stir thoroughly and leave to mature for 5-10 days.

Vermicompost Tea:
If you have been using a vermicomposter with a tap to drain excess water, then this can be used as an excellent tea mix. It is however quite strong and so a mix of around 40-50 parts water to one part tea is usually sufficient for most plants.

With regard to these 'teas' a ladle or cup of tea once or twice per week is usually sufficient.

Mulching

I am a keen advocate of mulching in general, for the reasons listed below.

1. Mulching keeps in the moisture content around the plant where it is most needed.

2. It suppresses weeds that would otherwise fight your plants for nutrients and water.

3. As the mulch rots it adds nutrients and humus into the soil, improving the soil condition and crop yield.

4. Though not really mulch – if you cover the soil with a weed suppressant fabric, this will also warm the soil slightly – good for early growth.

These attributes apply to vegetable growing in containers as well as in traditional vegetable beds.

Bark chippings are my favorite mulching material for my pathways between the vegetable beds, or indeed the planters I have around a particular area.

This is easy to walk on and prevents mud being spread around in the wetter weather, as well as acting as an excellent weed-suppressor.

For cucumbers, marrows etc I use just straw to mulch. This gives the vegetables a dry bed and discourages the dreaded slug.

That said, if you have a lawn that needs constant mowing (as they do) then a good idea is to use the lawn clippings as a mulch. Spread this about 2-3 inches thick between the veggies and allow to rot down to produce a nitrogen-rich feed for the plants.

For a more effective weed-suppressant, lay out some newspaper or cardboard before adding the lawn clippings.

However it has to be said that unlike planting in a garden or perhaps a raised bed; growing vegetables in planters usually means that there is not much space for mulching. However it is still worth-while, and perhaps more relevant when your planters are prone to dry out quickly.

Watering your planters

OK, I guess everyone knows that plants need watering or they will simply die. But what exactly does water do to a plant ? Three main things actually, they are:

Turgor, or rigidity. Water pressure within the stem of the plants creates Turgor so that a plant is able to stand.

Water enables the nutrients in the soil to energize the plant through the roots.

The process of photosynthesis means the plant uses light, carbon dioxide and water to make sugar.

As mentioned, planters are indeed prone to either drying out to quickly, or getting waterlogged by over-watering. Whether by nature or nurture-so to speak. Water logging blocks the oxygen source to the roots of the plant, and so the plant dies unless remedial action is taken in time.

Fungal diseases also thrive in wet conditions, making this a "double whammy" for the poor plant.

How do you know if your plants are over-watered ? Well the tell-tale signs that a plant is getting too much water are:

Leaves yellowing from the bottom up

Soil turning green

Grey mold appears on the plant

Plant has stopped growing

Plant is wilting badly

Prevention of this is simple. Make sure that you have prepared your planter properly as per the earlier instructions. Do not be to enthusiastic when it comes to watering, but even if you are, proper drainage should allow for the soil to reach a natural level.

Be observant ! just watch your plants for the signs of overwatering, and be ready to remedy the situation. If there is indeed fungal growth, then you may have to apply a fungicide to remedy it.

Signs of under watering include..

Dry, hard soil or compost

Plant leaves tend to go brown and crisp

Plant shrivels and dies !

You will notice the list here is shorter ? Fact is that many more plants die from overwatering than the opposite, largely because the overwatering starves the roots of oxygen and so the plant reacts faster in many cases.

Chapter 5: Matching Planter to Plant

It's difficult to think anything but pleasant thoughts while eating a homegrown tomato.~ **Lewis Grizzard**

Planting Tips

Now that you have your planters all sorted out, you should be ready to get them planted out with the vegetables or fruit, flowers even of your choice.

This is quite a simple process, but requires some careful thought. For instance you must decide not only what you would **like** to grow in your planter, but also what can be grown with a reasonable expectation of success.

For instance if you live in Scotland, you should not expect to grow peppers or sweet corn in a planter, unless it was inside a greenhouse or cold frame.

Also, it should be obvious that potatoes will not produce a good crop if your planter is only a few inches deep !

With that in mind, here are a few suggestions for planting vegetables or fruit in the different kind of planters.

Hanging baskets

Best suited for obvious reasons for short plants such as strawberries, however provided you have not placed the hanger too high, then you could consider using slightly longer chains as suggested earlier, and growing climbing strawberries or vegetables such as peas or beans up the chains.
A hanging basket can also be a good place to plant your lettuce as it is free from the predations of slugs and other ground insects.

Herbs as well grow vigorously in a hanging basket, which is just about the ideal size for a miniature herb garden.

Deep planters

Deep planters such as the potato planter made from tires or indeed a wooden sided box, suit other things besides potatoes of course. Into this category would come your root vegetables such as carrots, parsnips or sweet potato perhaps.

Other root vegetables such as beetroot, turnip, radish, celeriac etc, can also be grown here of course but strictly speaking can just as easily be grown in a much shallower planter.

Even a decent sized plant-pot as long as it is over say 10 inches deep can be used to grow carrots quite successfully, especially if you raise it up of the ground away from the carrot fly, which tends not to fly over two feet high or so.

Shallow planters

Anything under say 10 inches would be regarded as a shallow planter. However there is a good selection of vegetables that can be grown in these, the main problem is that the shallower the soil then the harder it is to keep the moisture content just right.

They are much more prone to getting water-logged or drying out very quickly. That said, even an ultra-shallow planter such as the Gutter planter, can grow a selection of herbs for instance such as parsley, mustard , sweet basil, sage thyme etc, and can also be used to bring on seedlings as mentioned earlier.

Wheelbarrow's and other odd-bods

Containers such as old wheelbarrows and other slightly more ungainly looking planters are ideal for growing climbing plants, such as peas or beans. Training them around the structure not only keeps the vegetable's well ventilated, but also makes something interesting and pleasant to look at.

A climbing strawberry plant such as the Mount Everest climbing strawberry can look quite spectacular, in this situation.

Your own imagination is perhaps the only limit on the kinds of plants you can grow, in the different planters available. Whether you are growing carrots in an old wellington boot, or potatoes in a child's discarded pram; the fact is that there is no end to the containers we can use to plant vegetables that will enhance our larder.

With the present move towards recycling and 'saving the planet' it is suddenly very much in fashion to do something that helps, instead of hinders our move to a safer cleaner environment.

As recycling has grown in popularity, so I am pleased to see is the idea of using material that would often be thrown into the rubbish tip, to produce great healthy vegetables straight from the garden with no "air miles" to count at all, and zero carbon footprint.

I sincerely hope that this work has been of some help to you, and has given you fresh ideas when it comes to planting your vegetables in containers of many kinds.

Happy Planting!

MY NOTES / TO-DO PAGE

MY NOTES / TO-DO PAGE

Book 5

Tomato Container Gardening: Growing Tomatoes In Containers, Planters & Other Small Spaces

The Easy Way To Grow Tomatoes In A Small Space

James Paris

Published By

www.deanburnpublications.com

Introduction

I vividly remember as a child of 11 years old, growing my first tomatoes. We had moved to a smallholding that had an old rickety Victorian greenhouse with no glass in it. I helped cover the greenhouse in clear polythene, and then was set the task of growing tomato plants for the first time ever. The results were fantastic, and I remember how proud I was to take the first of the tomato harvest into the kitchen to my mum.

The sweet full flavour of the fresh tomatoes was just enhanced by the fact that they were the work of my own hands, and from there on a love for growing tomatoes has remained with me.

But what if you do not have a large greenhouse – can you still grow tomatoes? Of course you can. You can grow prize tomatoes in a very limited space, even in an apartment building; it is a system called container gardening, and it works for many kinds of vegetables including of course tomatoes.

One of the great things about container gardening is that you do not have to be super-fit to do it. No digging up huge garden plots or wheel-barrowing heaps of dirt around, for this reason it is well suited to most physical abilities.

Age is of hardly any concern, which makes container gardening suitable for young and not-so-young alike. Ready to get started? Ok, lets look at tomatoes and their suitability for container gardening.

Fruit or Vegetable?

The humble tomato is probably the best known; most used vegetable in the world today; which is somewhat strange considering the fact that it is in fact a fruit! Technically speaking the tomato is a fruit, as true fruits are developed from the ovary of the flower which in turn contain the seeds of the plant.
The argument over whether or not it is a fruit or a vegetable is largely due to the differences between scientists and cooks, so it is not something to lose any sleep over in my opinion. Chefs generally regard a tomato as a vegetable because it is primarily used in savory dishes rather than in sweet dishes.

Whether you regard it as a fruit or as a vegetable, there is no doubt that the tomato is the most versatile of plants and is used massively in sauces, relishes , pickles etc; as well as innumerable different dishes.
In fact it is difficult to eat a savory dish in a restaurant today, that has not been flavored in some way by tomatoes or tomato extracts – which is bad news if you have an intolerance to tomatoes.

Back to the fruit versus vegetable issue – for the sake of continuity throughout this publication the tomato will be regarded as a vegetable, as I have always regarded it as such.

However if you want to call it a fruit – then I have no problem with that either!

Ok, now that is sorted out we can get down to business, and take a good look at the 'king of vegetables', and just how we can grow them to best advantage in containers of many shapes and sizes.

Why grow them in containers?

There are many reasons of course why you may wish to grow tomatoes in containers, but probably the most prevalent would be that you simply do not have the space to grow them otherwise. I was very fortunate to have grown up in a situation where we had a large old Victorian greenhouse, where I could grow tomatoes without any problem or real limitations with space.

Most people however do not have that luxury; does this mean that tomato growing is out for them? Definitely not, as this publication will seek to emphasize.

The growing of tomatoes in containers makes it possible for the individual with even a limited space, to grow tomatoes very successfully and enjoy their own fresh picked tomatoes, even if they stay in an apartment in New York City!

It may well be that you have a patio, but have no growing space as such. This also is an ideal situation for growing tomatoes in containers, as a patio is generally designed to catch the most of the sunshine, so suits tomato growing ideally.

Tomato plants can also be grown in containers for the same reason that you would grow flowers, or ornamental shrubs – they look great! This can be a real talking point with your guests as almost everyone is interested to some degree or other in what their host (or why) is growing in their containers. You will find that this can lead to discussion on growing all sorts of fruits and vegetables in containers; strawberries, chilies and runner beans – to name just a few.

One of the other reasons for growing your tomatoes in containers rather than in a traditional garden plot, is the lack of weeding needed. Since you have planted your tomato plant in a pot with good compost (hopefully!) there are no weeds to take care of apart from the ones that may blow in on the wind in seed form. This being the case, they are super-easy to remove.

Final reason for growing tomatoes in containers is that if you have a window sill or a patio, porch or balcony, that is exposed to a minimum of 6 hours full sunshine; then you can enjoy your very own sweet tasty tomatoes straight from the plant.

These are a far cry, from the tasteless bland tomatoes that are sold in most supermarkets today. This is the same with all home-grown vegetables; you just cannot beat that home grown flavor, partly due to the fact that it has not been lying on a supermarket shelf for who knows how long, and partly due to the feeling of self-satisfaction that comes with growing your own veg.

What You Need To Get Started

Head for the light!

As mentioned earlier, tomatoes need the sunshine – and lots of it. Experts will insist that tomato plants need 8 hours minimum of sunshine per day. Whilst I would agree that this is great in an ideal world, I have grown them in less than 6 hours sunshine per day quite successfully.

The bottom line though is that you want to try for 6 hours very minimum if you want to get a decent harvest of tomatoes. They will grow in less than this but the harvest will be extremely limited.

One of the advantages of container gardening is that often it is possible to move the containers around a little to catch the most of the sun's rays, especially as the season goes around and the suns position of course changes in relation to your chosen growing area. It is even possible to make a little platform with wheels on, where you can easily move it around to catch the sun in certain circumstances.

Choose your containers

There are many things that you can use for containers when it comes to growing your tomato plants, and probably the first thing to spring to mind is large plant pots. Something in the region of a five gallon pot or container is really needed for allowing the root growth required by the plant, to pull in the necessary nutrients for a good crop of tomatoes.

When choosing a pot, then go for something around 12 inches deep ideally, and try to go for a plastic or fiberglass design. The reason for this is that clay pots can dry out very quickly, which is terrible news for tomato production; and have a tendency to crack – which again is bad news.

Large clay pots can also be very heavy, which may be a consideration if you intend moving them around. Another good reason to choose plastic is simple economics; a cheap black plastic 5 gallon bucket or pail bought at the local home improvement store, is ideal for growing tomato plants and you can pick them up "cheap as chips". Perhaps not the most attractive of containers – but excellent for the job.

Timber planters can of course be used as well, but a good tip here is to line the inside of the container with plastic. This will help prevent it from drying out as the timber absorbs the moisture in the compost. Keep the base clear though and use gravel or broken clay pots as drainage, as you would normally. This method works quite well also if you do plant in an unglazed clay pot.

A raised bed garden, can also be looked upon as a kind of container, inasmuch as it contains the growing medium. This can be a great alternative if you want a more fixed solution to growing your tomatoes, and can vary in size to suit your requirements. It has all the advantages of any container, in that it requires only a fraction of the care that a traditional garden does but at the same time is capable of producing a superior crop of tomatoes.

Tomato grow-bags are again a popular 'container' for growing tomatoes. Preparation is minimal as the compost mix is already there and all that is needed is for you to cut out the top circles for your plants and get planting.

These bags, filled with growing compost, can be bought in several different sizes at just about every garden center; and indeed can be a great way to grow your tomatoes with minimum fuss.

Make sure that you loosen up the contents of the bag before you attempt to plant anything – and before you have cut the holes! After being flat-packed in storage for several months at the suppliers the compost material can be VERY compact when you first purchase the bag. Do this by simply laying the bag on its edge and crushing down on it – similar to fluffing up a bolster!

Tip here is to cut a few slits in the bag about half an inch or so up from the base to allow for drainage.

These are particularly handy if you are using a growhouse – which is basically a little miniature polythene greenhouse which has become increasingly popular amongst vegetable growers of all persuasions.

Hanging Baskets. While it's true that hanging baskets are particularly used to plant out trailing flowers that will hang attractively from the front porch – they can also be used very successfully for growing several tomato varieties.

The main thing to watch out for here is the basket drying out, as tomatoes do require a fair amount of water especially when they are fruiting. Counter this by adding a good amount of water retainer such as Pearlite to your compost mix.

You can see more on this tomato variety in later chapters.

Preparation

Drainage

Whilst this may seem obvious, it is very easily overlooked when preparing a container for planting. If you are using a black pail, then be sure to drill several holes in the base. Failure to do this first step will lead to your plants getting waterlogged in no time at all, which will lead to them rotting at the stem.

Next fill in about two inches of mixed gravel or broken down clay pots, to help with the overall drainage. If you are using clay or porcelain pots then these will usually have the drainage holes already in them; as will plastic plant pots.

Often though, other plastic containers require you to drill the holes yourself, which you can do using a 3/8 drill bit ideally. As mentioned earlier; if you are using bare clay or wooden containers then it is a good idea to line the sides with plastic to stop the timber from soaking in the water and drying out the compost. Tomatoes hate dry soil, so this would result in deterioration of your plants quite quickly if this were to happen.

Compost

Tomatoes are not demanding with regard to compost, and almost any of the compost varieties that you buy at your local garden centre will do. The main thing is that you do not fill your container with garden soil. This can get very compact in a container, and will get even more so as the season wears on. A mixture of compost and organic material mixed with well-rotted horse manure is excellent for tomato plants, and will help ensure a bumper crop.

One way to help prevent your container from drying out, is to add superabsorbent crystals to the compost mix. There are a variety of these that you will find in your local garden center or online. These will absorb the moisture when you water the plant, and will release it slowly into the surrounding compost, acting like an emergency water supply for your tomato plants.

Choosing Your Plants

When choosing tomato plants for your containers, you have to consider not just what variety of tomato you would like; but also what variety would suit the containers you have or indeed are about to buy. It is a bit like the chicken and egg scenario – which comes first. In the case of the tomato plant it is not quite so incomprehensible, thankfully!

Some tomato varieties are grown for the size of the fruit, or indeed the overall volume of the crop that you may expect from it. Others may be grown because it is the most successful variety to grow in your particular part of the world. Still others are grown according to the budget or the varieties readily available. However unless you intend to specialize in a particular variety, I do not think it pays to get overly complicated when it comes to choosing your tomato plants, provided a few general rules are followed, each of these following varieties can do very well in containers. The two main categories of tomato plants are known as determinate and indeterminate tomatoes, as explained in the following.

Determinate Tomatoes

Determinate tomatoes are generally known as bush tomatoes, and are probably the most commonly used for container gardening. Being of limited height, they will not topple over as easily as the indeterminate varieties. They will usually grow to an average height of between two and three feet, and usually do not require to be supported with canes or other supports. However I usually prefer to support all my tomatoes as if you have a bumper crop, then it is likely to topple the plant in the container. This variety of tomato tends to produce all its fruit at the same time over a period of about two weeks, after it has stopped growing at its terminal height, and then the plant will die off. Some good bush tomatoes would be:

Scarlet Champion: A sweet to slightly tangy flavor, easy to slice and a favorite for salads.

Oregon Spring: Lovely small tomato, specially adapted in Oregon for the cooler climate. This makes a tasty sweet and juicy fruit.

New Big Dwarf: A large tomato for the patio, reaching up to 1 pound per fruit. This is a meaty flavorsome tomato great for any dish, or sliced for a sandwich.

Patio Princess: Very tasty small tomato produces a good crop, and is ideal for the container gardener as two to three plants can be grown in the larger pots.

Dwarf Bush tomatoes

Dwarf bush are actually part of the determinate variety of tomato, and are also known as hanging basket tomatoes, for the fairly obvious reason that they are perfectly suited to growing in hanging baskets. The main thing here is to make sure that you do not have your hanging basket so high as to make it difficult to tend to your tomato plant or harvest your crop.

Tomatoes in hanging baskets do have a tendency to dry out quickly, so be sure to cover the points listed in the previous chapters and prevent this happening. One of the benefits of growing this variety in a hanging basket, is that they are not so vulnerable to slugs. This can be a real problem normally because the lower clusters tend to lay on the ground, making them vulnerable to this garden pest.

Some good tomatoes for the hanging basket include:

Tiny Tim: Produces a small cherry tomato, great for salads and snacks. Perfect for hanging baskets.

Tumbler: Ideally suited to a hanging basket due to the growth pattern of spreading out and 'tumbling' over the side of the basket. Small fruits just over 1 inch in diameter but full of flavor.

Garden Pearl: Produces a good yield of elongated fruits that hang well over the sides of the basket; deep red and juicy tomato.

Hundreds and Thousands: Fantastic plant for a high yield of cherry-small tomatoes bursting with flavor.

Indeterminate Tomatoes

Indeterminate tomatoes, known also as vine tomatoes, grow throughout the season, up to a height of ten feet or so, if allowed to do so. This variety has to be supported by canes or string, otherwise it will just fall over with the weight of the tomatoes. It produces tomatoes throughout the season, all ripening at different times, and will eventually die off with the first frost.

This variety needs a little more care in that the side shoots have to be removed, otherwise it will soon get out of control and you will have more foliage than actual tomatoes. Many commercial tomato growers in fact remove almost all of the foliage in order for the plant to concentrate on growing the actual tomatoes, and have a single stem carrying the clusters of tomatoes.

Most growers will say not to grow this type as a container tomato, however I would just add that if your container is well enough secured and not likely to be moved, then this is an option. Provided that you have a good enough support in the form of a cage or perhaps growing the actual vine up a fence or wall, there is no reason why you cannot grow vine tomatoes in a container.

Most popular vine tomatoes include

Mortgage Lifter: This was grown by crossing four large tomato plants, by Charlie Byles. It was Called the mortgage Lifter as he was able to pay off his mortgage with the results!

Giant Pink Belgium: This plant produces a fruit that can weigh anything from two to five pounds! Needless to say it makes this plant a favorite for sandwiches.

Big Rainbow: Another plant producing large tomatoes grows up to eight feet tall and you can save the seeds for next years crop.

Goliath: As you may guess from the name, this plant produces giant fruits over 1 pound in weight. Very disease resistant, it produces a great tasty tomato.

Heirloom Tomatoes

Heirloom tomatoes, sometimes called heritage tomatoes, are simply older varieties that are tried and tested to 'produce the goods'. Tomato developing is an on-growing process especially with the commercial growers, and so for various reasons some varieties get left behind, but not forgotten by their advocates. These tomatoes tend to be 'old favorites' in the world of tomato growing.

Some popular varieties of heirloom tomatoes include:

Sub Artic Plenty: This is a determinate variety, and one of the earliest crops; producing a good crop of tomatoes about 4 oz in weight.

Green Sausage: One of the 'oddball' tomato types. It is a long green fruit with yellow stripes. A tasty tomato with a real talking point aspect to it for any dinner party!

Brandywine: One of the slower growing heirlooms, this is nevertheless a popular tomato with a rich sweet flavor.

Speckled Roman: A thick meaty tomato, excellent for sandwiches. Great for canning or sauce making.

Each of these varieties will produce a tasty tomato fit to grace any dining table.

Planting Your Tomatoes

When it comes to planting out your tomatoes, no matter what the variety or type there is two things absolutely fundamental to a successful crop of tomatoes; plenty of sun (at least 6 hours) and adequate water. Failure in either department will result in a poor quality crop of tomatoes.

Tomatoes prefer a light loamy compost with plenty of organic matter. A couple of handfuls of a slow release organic fertilizer mixed through the soil will also help boost that early growth.

I always prefer planting actual seedlings rather than grow the plants from seed, basically because it is much quicker to do so, and your plants get off to a good start. You can of course grow from seed with most of the heritage especially, but it can be a slow process and needs a lot more preparation with heated cold-frames etc before you can get down to planting.

The main advantage that planting straight from seed has however, is that you have a far greater range of plants to choose from, as you are not just restricted to whatever your supplier has in stock.

The process is the same for all of the tomatoes no matter what your tomato container is. Simply make a good sized hole in the compost and plant deep, so that the compost is above the base of the plant. You will notice small hairs near the base of the stem? These should be covered and will grow into extra roots that will help the plant get established that much faster. Water vigorously to make sure the plants are well tamped in. Make sure that the last frosts of the season have passed before planting out!

If planting in a raised bed 'container', then plant out about 18-24 inches apart. If in grow-bags then a little closer will be fine, say about 10-12 inches for most grow-bags. Pots of course will only take one or at most two plants unless you are planting some of the Determinate or Dwarf varieties.

As mentioned earlier, ensure that your container is prepared properly and allowance has been made for drainage, and the container properly placed to get the best advantage of the sunshine.

Support

Some of the bush varieties of plants will not need any support, although if you have a heavy crop it is always better to support the plant rather than have the extra strain on the stem, perhaps leading to damage to the fruit as well as the plant itself.

The Indeterminate or vine tomatoes certainly do need support, even if it is just a length of tight string, otherwise they will fall over and get damaged. Depending on the kind of container that you have chosen in which to grow them, tomatoes can be grown within a cage system. This is simply a roll of 3 or 4 inch wire mesh staked around the plant with canes, in order for the plant to grow up within the cage and thereby be supported.

Most of the time however, I like to support with a single tall cane, firmly planted in the ground. This allows more space to harvest the crop and also to tend the plant. If you are planting a tall growing vine tomato in a pot of some kind, then the cane or frame will most likely have to be tied in turn to a nearby fence or wall. This will stop the whole lot from falling over, especially if it gets a little windy.

Feeding Time

Once the first flowers have opened, then it is time to start with a slow feeding regime. To begin with I would just feed about once per week with one of the common liquid fertilizers such as Tomorite.

If however you want to try organic, I have had great success with a slurry made up from horse manure and water, all mixed up in a 45 gallon drum! Not everyone's idea of fun perhaps, but I have got some great tomatoes from it. Simply add a couple of pails of manure to the drum of water and mix thoroughly.

Once tomatoes have started to form, then I would increase feeding to about twice a week with one ladle of the manure mix to the base of each tomato plant. Sticking to the organic theme, liquid seaweed, bone meal or fishmeal, all make good fertilizers.

Using the liquid Tomorite, or any other chemical fertilizers, be sure to read the instructions and not over-feed the plants.

General Plant Care

Pruning

Tomato plants, like most plants, need regular pruning or other care if you are to get the best out of them. Fortunately with the tomato plant this is quite a simple affair. With the vine tomatoes particularly, they tend to send out side shoots or suckers between the stem and the leaf.

These shoots have to be trimmed away otherwise the precious sugar that should be going to the fruit is diverted to these growing shoots. Leaving these shoots to grow will mean that the plant will get over-loaded and the crop will be poor. Be careful when doing this, that you do not cut away the fruit bearing branch by mistake.

A top tip here. If you like you can let one of these shoots grow out for about 6 inches or so, then cut off carefully and plant in some wet compost. This will produce another plant that can be used elsewhere or as a replacement in case of disaster.

Once the plant has grown to full height or about 6 trusses from the ground you can trim the top and halt the upward growth. This will allow the nutrients to be used to feed the tomatoes instead of going to encourage leaf growth. In fact the commercial growers tend to strip away the leaves more than half way up the stem when the plant is full grown, in order to get a better tomato crop. Trimming away the leaves like this also helps stop or prevent tomato blight on the leaves.

The bush tomatoes generally do not need this kind of pruning as they are developed to grow bush-like and have a shortened growing and fruit-bearing season. Any signs of tomato blight however can be treated the same way, by stripping of the affected leaves, and removing them from the scene.

Pest Control

One of the advantages of container gardening it has to be said, is the fact that you are a few steps ahead of the traditional garden, when it comes to pest control; as shall be seen in the following writings.

Controlling pests is an on-going task for any gardener, whether it is vegetables, flowers or fruit; they all need to be taken care off. There are however different types of pests, for instance it can be the flying kind, the creepy crawling kind, the four-legged kind, or just plain old pestilence !

Fortunately however, it is not as bad as it seems, and a few simple precautions and preparations will usually ensure that you are 'ahead of the game' so to speak. The kinds of attacks you may have against your precious tomatoes are subject to many different influences, mainly the weather and the area in which you live. Here are a few of the more common afflictions and how to deal with or prepare for them to give you the best chance of success.

Rabbits

Usually I would give a long-winded explanation regarding the predations of rabbits on a vegetable patch at this point; as I have suffered to my cost over the years with these menaces! However since this is about container gardening, and as such not so vulnerable to rabbits; I can say through experience that a fence around your containers, or the area that you are planting in, is the best deterrent. It has to be made with garden mesh no more than

1/1/2 inch (to keep out the young ones) and about two feet high is usually sufficient.

Ideally the mesh should go down into the ground about 6 inches (150mm) or so on the outside, with any excess at the bottom of the trench flowing out in an L shape away from the fence. This will stop the rabbits if they do burrow down, from getting through the mesh at the bottom of the trench.

Birds

It has to be said that birds are possibly the easiest pest to control around the garden-though they can be very persistent! Simply apply some nylon netting around your planters and that's it sorted. Raised bed gardening is particularly easy for this method, as the mesh can be put around a frame secured to the raised bed.

This can then be lifted up to gain access to the vegetables underneath. Whatever the situation, it is important to stress that you must make sure the netting does not actually touch or come close to the plants, particularly strawberries as the birds will just peck through to them.

Sometimes in vegetable patches you will see strings with all sorts of foil or paper attached, weaving over the veg patch. This is ok for young seedlings, but not ideal. I have witnessed many times the birds just ignoring this, as they will do with any constant motion that they get used to.

Scarecrows now….not really, they tend to be fine for a short period, again until the birds get used to them and then ignore them completely.

It is perhaps important to point out here as well that birds are not always bad news for the vegetable gardener, they can be a real asset when it comes to picking off caterpillars for instance. Snails are a favorite for the song-thrush, and all sorts of creeping things that aim to do you mischief, are on the birds menu!

In fact apart from the damage they can do your fruit plants, or your early seedlings, I regard birds as doing more good than harm – most of the time! Pigeons on the other hand can wreak havoc on your cabbages, so have to be dealt with, usually with netting or bird scarers of some kind.

Slugs

I've yet to meet a gardener who likes slugs! It's still one of God's creatures fair enough, but one that I can do without. Slugs can decimate most vegetable types, but especially brassicas like cabbages or lettuce plants, that overnight can be ruined by these pests.

Fortunately again, container gardening is especially easy with regard to slug control, as the plants are up off the ground to begin with. The usual organic method is a simple beer trap (a jar with beer inside) sunk into the ground and covered with a tile, leaving room for the slugs to get under of course.

These then have to be removed in the morning – quite disgusting, but effective. However for outside containers, it is usually simply a matter of either surrounding your container with a copper wire or tape (slugs will not cross over copper due to a chemical reaction with their slime) or sprinkling some salt or a salt solution around the outside base of the container.

There is also of course the chemical solution via slug pellets that are extremely effective, but deadly to many things besides slugs. Imagine for a moment that the gardener's best friend the hedgehog, comes and eats slugs killed with slug pellets. The hedgehog is dead-simple as that, and your best natural pest controller is no more! The same can be said for any other creature that eats the dead or infected slugs, particularly frogs and toads. for that reason I use slug pellets only in dire circumstances.

Others

In this category would be mice for instance. Though not usually a great problem, they do like to snack on your bean shoots or seedlings – as my wife discovered just recently when she went out to inspect her young beans - The blessed mice had done for almost half of them! The answer here is to cover them with fleece until they are a bit more established. If germinating in a greenhouse or cold-frame then trapping may be the answer, if the problem is severe.

Moles and gophers, depending on where you live may also be a problem. Little can be done here except trapping or chasing with sonic devices perhaps. These can work against all sorts of critters including mice. With a raised bed garden this is not so much of a problem if you have built it in accordance with the example in my raised bed book, with wire mesh covering the base before adding the compost.

Plague and Pestilence and creepy crawlies

The bad news is that there are about 30 different diseases that can come against your tomato plants; the good news is that they are unlikely to come all at once! Most tomato disease can be avoided by providing adequate ventilation, cleaning away infected plants or leaves, and keeping the area around your plants free from rubbish that is likely to harbor insects and infection.

With that said however, here are listed a few of the more common ones that you are likely to encounter at some time or other.

Tomato Blight: This is usually shown by the leaves of the plant going brown yellow and falling off, eventually leads to the whole plant rotting away. Usually prevalent in tomatoes grown outside, and in periods of heavy rainfall or overly humid conditions. Lack of ventilation between crowded plants can also be a factor.

If rainfall is the culprit and they are grown outside then there is little that can be done about it I'm afraid. Even with the chemical sprays available, if the conditions are right for it, then it is difficult to stop the spread. If you catch the disease early enough then sometimes just removing the infected plants from the site and all infected leaves of the remaining ones, may just do the trick. Do not add these plants to the composter, but instead burn them or remove them entirely from the area.

Grey Mold: As the name suggests, this is a light grey mold that may form on the plant, especially in a poorly vented greenhouse. It usually starts with a wound on the plant of the kind caused by removing side shoots, and eventually leads to a patch of grey fur growing and becoming black on the plant.

This will affect all parts of the plant including the fruit. The best way to stop this is to make sure that there is adequate ventilation and free flow of air around the plants themselves. Treating the soil at the end of the growing season with Jeyes fluid or something similar will help kill the spores that may have been left behind in the soil.

Whitefly: This can be a real pest amongst tomato growers, especially in greenhouses as it lays eggs on the underside of the leaves. They also excrete a sticky substance which in turn leads to sooty mold growth. This is what makes the whitefly a notorious vector for all sorts of plant disease.

The effect to the plant is predictable in that the plant will weaken and die in extreme cases. One of the ways to help protect against infestation is to plant French marigolds around your tomato beds as the whitefly is not attracted so much by the smell of the tomato plant – which is masked by the pungent smell of the marigolds.

Infestation with this pest means that all the fruit has to be thoroughly washed, if it has not been damaged and is still edible. Plants at the end of the season are best burned before disposal.

Aphids or Plant lice: This is a most common type of infestation, caused by greenfly, blackfly or whitefly. These insects multiply vigorously, and in no time at all you can have a major infestation on your hands. The problem is not so much the damage that these sucking plants cause by themselves, but rather the diseases that can be caused by the sticky substance they leave behind.

An effective natural control against aphids is the lady bug or lady bird beetle, which feeds voraciously on the aphids. Another control method which I have used quite successfully is a mixture of washing up liquid and water to spray them with. As the aphid breaths through the skin, this has the effect of blocking the pores and killing the aphids. Just do not make the mixture to strong, and test on one part of a plant first.

Tomato fruit-worm: This is actually the larva of the moth *(Helicoverpa armigera Hübner)*and can do immense damage to the plant as the caterpillar grows big and strong; strong enough in fact to attack other caterpillars and even bite!

The larva infest the stems, leaves and the tomato fruit making it inedible. First signs of this pest are the white eggs laid near the topmost flower stems. When the eggs hatch, the larva burrow up the stems and into the fruit, causing infestation. Spraying with a suitable insecticide should begin with the first sightings of these eggs. Badly infected plants should be removed and burned.

Tomato Hornworm: This section would not be complete without the addition of the largest caterpillar you are likely to see in your garden – if you are unlucky! Growing to a massive 4 inches in length when adult, this caterpillar can consume a whole plant virtually overnight, as it has a voracious appetite.

The caterpillar is the larva stage of the hawk or sphinx moth, also known as the humming bird moth. It lays greenish white eggs on the underside of the leaves, which hatch four to five days later. The caterpillar grows to full size in only about 4 weeks.

The best way of treating this infestation is to remove the leaves if there is eggs attached. The caterpillar itself is so big it is easy to spot, and can be picked of by hand and disposed off. If you see a caterpillar with egg sacs on its back, then leave it if possible. These are the eggs of the parasitic Braconid wasp and will hatch out and form an army of defenders for you.

Summary

I would finish by say that if you are new to tomato growing or vegetable growing in general, please do not let this list of possible plagues and pestilences put you off growing your own. The fact is that you are highly unlikely to get any major infestation unless you are extremely unlucky, if you follow the general advice on ventilation and cleanliness.

If you do then put it down to experience (or lack off) and deal with it. Whether you are growing vegetables or rearing livestock, you will have your casualties. Your challenge as a gardener is to limit them to the best of your ability, and concentrate on 'coming up with the goods' as they say.

Gardening should not be a chore; though it can be hard work at times. Apart from providing fresh vegetables for the family, gardening can be excellent for 'me time'. A chance to get away from the hussle and bustle of life; to forget the worries of the office or workplace.

Tending the garden can have a real therapeutic effect on the spirit, as well as a healthy effect on the body. It has great social implications as well as financial ones; I would recommend it to everyone.

Authors Note

Finally - a huge thanks for purchasing this book, your support is appreciated. If you can spare the time, and have the inclination, a good honest review on Amazon would also be much appreciated. Thanks again.

James

For further reading, you can see the full range of books that I have on offer on my Amazon author page.
www.amazon.com/author/jamesparis

Relevant Books by Same Author

Raised Bed Gardening 5 Book Bundle

Raised Bed Gardening 3 Book Bundle

Companion Planting

Square Foot Gardening

Square Foot Vs Raised Bed Gardening

Vegetable Gardening Basics

Companion Planting

Straw Bale Gardening

Blog: http://planterspost.com

MY NOTES / TO-DO PAGE

MY NOTES / TO-DO PAGE

Printed in Great Britain
by Amazon